PRETTY PACKAGES

45 Creative Gift-Wrapping Projects

SALLY J SHIM

photographs by MADELENE FARIN

CHRONICLE BOOKS
SAN FRANCISCO

for

SANG, JEREMIAH, and JUDAH

Library of Congress Cataloging-in-Publication Data available.

ISBN 978-1-4521-2599-2

Manufactured in China

Designed by Hillary Caudle
Photographs by Madelene Farin

10 9 8 7 6 5 4 3 2 1

Chronicle Books LLC
680 Second Street
San Francisco, California 94107
www.chroniclebooks.com

CONTENTS

INTRODUCTION
4

MATERIALS AND TOOLS
6

GIFT WRAP TECHNIQUES
10

PROJECTS PART 1
WRAPPING PAPER

Hand-Carved Stamped Wrapping Paper **25**
Photograph Wrapping Paper **27**
Stitched Wrapping Paper **28**
Patterned Wrapping Paper **31**
Graph Paper Gift Wrap **32**
Dotty Wood Wrapping Paper **33**
Handmade Glitter Tape Gift Wrap **34**
Vellum Stickers Wrapping Paper **37**
Label Stickers Gift Wrap **38**
Letter Stickers Gift Wrap **39**
Polka-Dot Stickers Gift Wrap **40**
Painted Dots Gift Wrap **41**
Watercolor Wrapping Paper **42**
Hand-Painted Furoshiki **45**

PROJECTS PART 2
GIFT BOXES, BAGS, AND TOPPERS

Custom Cardstock Gift Box **51**
Felt Gift Box **53**
Washi Tape–Wrapped Tin **55**
Custom Drawstring Bag **57**
Personalized Fabric Bag **61**
Washi Tape Glassine Bag **63**
Stitched Felt Pouch **65**
Yarn Pom-Pom Topper **69**
Felt Letter Gift Topper **71**
Fabric Flag Gift Topper **73**
Wood Veneer Bow Gift Topper **75**
Fabric Button Gift Topper **78**
Tissue Paper Flower Gift Topper **81**
Paper Rosette Gift Topper **83**
Stitched Paper Garland **86**
Felted Ball Garland **89**
Brownie Mix Glass Jar Kit **91**
Baked Goods Packaging **94**

PROJECTS PART 3
GIFT TAGS AND RIBBON

DIY Gift Tags **98**
Confetti Gift Tag **100**
Paint Chip Gift Tag **102**
Wood Gift Tag **104**
Personalized Leather Gift Tag **106**
Clay Gift Tag **109**
Handmade Patterned Washi Tape **111**
Hand-Painted Patterned Twill Tape **112**
Hand-Dyed Twill Tape **115**
Painted Wooden Bead Ribbon Ends **117**
Linen Ribbon **118**
Custom Paper Envelope **120**
Nail Polished Brads **122**

RESOURCES
125

ACKNOWLEDGMENTS
127

INTRODUCTION

Making things and being inventive was a part of my upbringing. My father was an academic by day and a self-taught architect, carpenter, and handyman by night. My mother sewed our clothes when we were younger, and my sister and I have the photographs of all our matching outfits to prove it. Little did I know that several decades later I would be doing the same thing with my two sons.

While living as an academic, I started to share my creative endeavors on my personal blog. Blogging for me was a creative outlet, and much to my surprise, I became part of an extensive online community of creative women. Through my blog, I began to realize my creative hobbies could become a business, and I soon found great pleasure in making one-of-a-kind items by hand and selling them to people and to retail stores. One of my favorite things about selling my handmade goods, many of which were paper items, was finding creative ways to package the orders. Fueled by my love of distinctive packaging and thoughtful branding, I spent hours thinking of ways to package each product. Several years later, I started teaching packaging classes with fellow designer Joke Vande Gaer of Tokketok. Though I've taken a break from selling my own goods, I continue to blog and work on creative projects in my home studio in Busan, South Korea.

In a time when we rely on instant communication via e-mails, text messages, and tweets, it is good to slow down and put meaning into our actions. I admit I have a slight obsession when it comes to packaging—I love pretty packages and believe the packaging of a gift is an extension of the gift itself. What you see on the outside is a prelude to what is inside. I have been known to buy a product based on the packaging alone, and I always take note of how products or gifts are packaged. In almost everything in life, I believe presentation is important.

This book is filled with simple projects that can easily be completed no matter your skill level. Many take just a few minutes to complete! Each project also includes additional "bright ideas"—small variations for further customization or using alternate materials you may already have on hand. As you go through the book, feel free to mix and match projects and bright ideas to create the perfect package for any occasion—there's no need to wrap them as I've done in the photos. Try adding a clay gift tag to your watercolor gift wrap or a simple wood veneer bow to stitched gift wrap. You can hand-stamp fabric and use hand-dyed twill tape to make a custom drawstring bag. The combinations are endless.

I don't believe gift wrapping has to be fancy; in fact, I prefer simple packaging. But techniques and ideas also depend on how much time you have. For most of the projects in this book, you don't have to spend a lot of time wrapping your gift, but with a few simple steps, a gift can go from plain to special. If I have a little more time, I might opt to sew a personalized drawstring bag or pouch that becomes part of the gift.

I favor minimal and modern packaging and often try to make use of materials I have in my studio. There's no need to buy fancy tools and materials you'll only use once. Having a stash of supplies on hand will make it easier for you to create a compelling package. In my studio, I have a wide array of paper and craft supplies that I can use to wrap a gift. Visit your local craft or art supply shop for basic packaging items, but for inspiration, also drop by fabric, yarn, party supply, floral supply, office, or paper stores. I love to visit hardware and woodworking stores as well as restaurant and baking supply stores to find interesting and unexpected materials perfect for wrapping gifts. And if you don't live close to any of these stores, don't fret, because a wide selection of gift-wrapping supplies is available online. (I've shared my favorite shops and sources in the Resources section at the back of the book.)

And remember, you can find inspiration all around you. I discover ideas in many places, from the natural environment to my sons' artwork to images in magazines. I don't believe you have to live in the most beautiful city or visit museums to find inspiration (although it doesn't hurt). Take a walk outside, visit your library, go to a café and people-watch, or go online and peruse some intriguing websites. I am inspired every day.

Whether you're wrapping up a gift for a friend or packaging a custom order of your handmade goods, I hope this book provides you with packaging inspiration and also inspires you to play with simple materials in new ways.

MATERIALS AND TOOLS

The best thing about making pretty packages is that you probably already have a lot of the materials you need! If you have a stash of papers, fabric, ribbons, and basic gift-wrapping supplies, you will find that creating pretty packages is enjoyable and easy. I recommend having the following basic materials in your crafting closet.

MATERIALS

BAGS

Bags can come in a variety of fabrics and materials, such as muslin, paper, burlap, and glassine (thin, translucent paper). Depending on the size and shape of your gift, as well as how you'd like the overall gift wrap to look, you can choose to work with any of these types of bags. You can also make your own bags from fabric (see page 57) or paper (see page 14).

BOXES

Boxes are perfect for packaging gifts. In this book, you'll learn how to make your own cardstock boxes (see page 51) and felt boxes (see page 53).

DECORATIVE MATERIALS

Decorative materials add a special element to the gift. Apply them directly on gift wrap or use them on a gift tag.

PAINT: Acrylic, watercolor, and fabric paints come in a wide range of colors. You can purchase them in tubes or containers as well as pen form. Use paint to make patterns on gift wrap or twill tape, write a personal message or recipient's initials on a wood gift tag, or personalize a drawstring bag.

PENS: Pens come in a huge variety of colors and types (gel, ballpoint, paint, metallic, glitter). You can use them to write on gift tags or to create patterns on paper.

RUBBER STAMPS and STAMP PADS: You can find a wide variety of rubber stamps at arts and crafts stores and stationers. A creative alternative to a store-bought stamp is to have a custom stamp made. You can provide the design or work with a designer to come up with a design. You can also hand-carve your own stamps (see page 25). Make sure to use archival quality ink stamp pads for paper craft projects.

WASHI TAPE: Washi tape is a decorative paper tape from Japan, and is one of the most versatile craft supplies. It can be ordered online in a variety of colors, patterns, and widths. Use it to embellish wrapping paper, gift tags, cards, and boxes.

FABRIC

Fabric is ideal for wrapping odd-sized gifts. You can use fabric to make drawstring bags or a Japanese-style wrapping cloth, or *furoshiki*. If you only need small quantities of fabric to make drawstring bags or fabric tape, you can purchase small pieces of cloth called "fat quarters" at a fabric shop. For wrapping up packages, check out flea markets and vintage stores for beautiful vintage textiles. You can recycle clothing or use vintage tea towels and scarves to wrap up your packages.

PAPER

Many of the projects in this book call for different kinds of paper, depending on what you're making. Here's a list of the most common ones used.

BUTCHER PAPER: A type of kraft paper that can be used as an alternative to wrapping paper. It commonly comes in white but is also available in colors, and it comes in rolls. You can find a selection at party supply stores.

CARDSTOCK or COVER-WEIGHT PAPER: Heavyweight paper for handmade cards, gift tags, and gift boxes. It comes in a variety of colors, weights, textures, and patterns. Scrapbook paper is in this category.

CHIPBOARD: Thick, stiff board usually made from recycled paper. It can be used for making gift tags and gift boxes and comes in many colors.

DECORATIVE PAPER: Specialty papers such as Japanese yuzen paper and handmade papers can be found in art supply, paper supply, and stationery stores.

GRAPH PAPER: Graph paper or graphing paper is paper printed with fine lines that make up a grid. It comes in individual sheets or bound in notebooks.

KRAFT PAPER: A sturdy paper made from wood pulp, traditionally used to manufacture paper bags for grocery stores. It is an inexpensive alternative to wrapping paper. It can be purchased by the roll or in sheets, and comes in a variety of shades and weights.

ORIGAMI PAPER: Lightweight paper that is perfect for wrapping mini boxes. You can also tape several pieces of origami paper together on the wrong side and use it to wrap a small- to medium-size box. It comes in a variety of colors and prints.

RECYCLED PAPER: Old newspapers, calendars, maps, wallpaper, and book pages can provide a vintage feel when used to wrap gifts.

STICKER PAPER: Paper with a matte side and a peel-off sticky back. Make simple labels by printing words or pictures on sticker paper and cutting them out. You can also purchase patterned paper PDFs online and transform them into labels or stickers to seal bags or boxes. Sticker paper comes in a variety of colors.

TEXT-WEIGHT PAPER: Paper for making gift toppers and embellishments, such as rosettes, confetti, and gift bags. It comes in many colors and textures.

WRAPPING PAPER: A variety of colors, prints, and textures can be purchased by the roll or in sheets.

RIBBONS and TRIMS

Ribbons and trims can be used to tie gift tags on packages, secure the wrapping, and provide a decorative element like a bow on the wrapped gift.

RIBBON: Cloth ribbon is ideal for tying bows on gifts. It comes in a variety of materials and textures, such as grosgrain, satin, silk, and linen, and can be purchased by the spool or in smaller quantities at fabric or notions shops.

STRING: Made of twisted threads of cotton or other material, string can be used to tie gift tags on packages. String or embroidery and crochet thread comes in spools or small skeins and is found at fabric stores in a variety of widths, weights, and colors.

TWILL TAPE: A flat woven ribbon made from cotton or linen. Twill tape can be used to tie bows on packages and is ideal for dyeing or painting. It comes in a variety of widths, textures, and colors.

TWINE: Twine is ideal for packaging food items and attaching gift tags to packages. Find plain kitchen twine at the grocery store. Baker's twine comes in a wide array of colors.

TOOLS

Once you have the essential gift-wrapping materials, you need to add some basic tools to your craft toolbox.

ADHESIVES

When you are working on paper craft projects, you need to have a good supply of adhesives.

CRAFT GLUE: A traditional water-based white glue that dries clear.

DOUBLE-SIDED TAPE: Clear tape with two sticky sides. It adheres two pieces of paper together. It's great for gift wrapping so you do not see any tape on the package.

DOUBLE-SIDED TAPE GUN: A double-sided tape dispenser is one of my favorite adhesive tools. With one easy swipe, it applies a thin strip of double-sided adhesive on a surface. I recommend the 3M Scotch Adhesive Transfer Tape ATG 700 Dispenser, which lays down double-sided adhesive that is stronger than conventional double-sided tape.

GLUE STICK: A solid but sticky glue in a tube. It dries clear and will not wrinkle the paper.

HOT GLUE STICK and HOT GLUE GUN: A solid glue that comes in a stick form. When placed in the hot glue gun, the glue melts. It's ideal for using on materials that require a strong hold.

TRANSPARENT TAPE: Transparent tape for sticking paper together, such as Scotch tape. It is not truly clear, so you will see it on whatever surface you put it on.

XYRON STICKER MACHINE: My favorite sticker machine to apply adhesive to the back of paper, fabric, or any flat materials. It is available in many sizes.

BONE FOLDER

A bone folder is a multifunction tool that is essential when working with paper crafts. It is used for creasing and for scoring (creating indented lines on the paper), which helps you to fold paper neatly.

CUTTING TOOLS

It is important to have sharp cutting tools designated specifically for use with paper and separate ones to use for fabric so the blades stay sharp.

CORNER ROUNDER: A paper punch that rounds the corner of the paper.

PAPER CUTTER: Although you can cut everything with scissors, for making precise cuts, it is a good idea to have a paper cutter. This device also allows you to cut through multiple sheets of paper at a time. I recommend purchasing one with a self-sharpening blade.

PAPER PUNCHES: A paper punch is a handheld tool used to cut out shapes. They come in a variety of sizes, shapes, and designs and are perfect for making confetti, gift tags, or decorative elements on your package.

ROTARY CUTTER: Although traditionally used for fabric, I also use a rotary cutter for cutting large sheets of paper. I have two rotary cutters, one for fabric and the other for paper. You can use a straight blade or decorative-edge blade on the rotary cutter.

SCISSORS: In addition to standard-size paper scissors and fabric scissors, you should also have small paper scissors and small fabric scissors with sharp tips for cutting out small details. You may also want to invest in scallop or pinking fabric scissors to create scallops and zigzag edges on fabric. Decorative-edge scissors for paper come in a variety of designs.

SELF-HEALING CUTTING MAT: A cutting mat is an ideal surface for cutting with an X-ACTO knife or rotary cutter. It protects your work surface and keeps the sharp blade from getting dull. I use a large cutting mat for fabric and a small cutting mat for paper.

SPEEDBALL LINOLEUM CUTTER HANDLE and CUTTERS: A knife with a short blade used for cutting linoleum or carving blocks. The cutters come in different sizes and shapes and can be screwed onto the cutter handle.

X-ACTO KNIFE: A very sharp knife cased in a pen-like body with a replaceable blade. Use an X-ACTO knife with caution. It works for making crisp, intricate cuts (such as for the Personalized Fabric Bag project, page 61).

LARGE CLEAR RULER WITH A GRID and/or A T-SQUARE

A large clear ruler helps you make straight cuts on fabric and paper. A T-square also helps you to cut straight lines on wrapping paper.

LEATHER HOLE PUNCH and MALLET

You can create a hole in leather using a buttonhole leather punch and a plastic mallet (which are ideal when working with leather). Use the leather hole punch and mallet on top of a cutting mat. Punches are available in a variety of hole sizes. The mallet can also be used with leather alphabet stamps.

METAL RULER

A metal ruler with a cork back to prevent slipping is ideal for use with a craft knife.

GIFT WRAP TECHNIQUES

Now that you have the essential materials and tools on hand, let's review some basic gift-wrapping options.

DECIDING HOW TO WRAP YOUR GIFT

When deciding how to wrap a gift, first survey the size of the present. Mini to medium-size presents can most often fit inside a gift bag or gift box and be wrapped with paper. Large to oversize or oddly shaped presents might require a little more creativity. For odd-shaped gifts, try a furoshiki, a traditional Japanese wrapping cloth (see page 45), or create a custom gift bag out of paper (see page 14).

Wrapping food gifts can be as simple as using a base paper, such as parchment, freezer, or wax paper, and sealing it with a piece of washi tape, ribbon, or twine. You can also package small treats in wax paper bags, glassine bags, tins, glass jars, metal containers, paper boxes, or bags lined with parchment paper. You can find plain food packaging boxes at restaurant supply stores. And don't forget to add a personal touch with a recipe gift tag or card.

WRAPPING YOUR GIFT

When wrapping your gift, you'll want to think about what supplies to use and how much material you'll need. Here's a review of gift wrapping supply amounts and basic techniques.

BOX, GIFT WRAP, and RIBBON AMOUNTS

If you've decided to wrap your gift with a box and wrapping paper, you will have to determine how much gift wrap and ribbon you will need to wrap the box. In general, you want the length of the wrapping paper to be equal to the width of all four sides of the box, plus an extra 3 in/7.5 cm to overlap the edge. With the box centered on the paper, the width of the wrapping paper should extend past the edge on both sides of the box, three-fourths the depth of the box. For example, for a box that measures 5 by 5 by 2½ in/12 by 12 by 6 cm, you'll need wrapping paper that measures 18 by 8¾ in/46 by 22 cm. In general, you will need a piece of ribbon that wraps around the length of the box twice.

Here's a quick guide to use as you work through the projects in this book to see how much wrapping paper and ribbon you'll need:

MINI BOXES
(1–2 by 4 in/2.5–5 by 10 cm):
WRAPPING PAPER: 1 sheet, 6 by 6 in/15 by 15 cm
RIBBON: 12 in/30.5 cm

SMALL BOXES
(3–4 by 5–6 in/7.5–10 by 12–15 cm):
WRAPPING PAPER: 1 sheet, 10 by 10 in/25 by 25 cm
RIBBON: 18 in/46 cm

MEDIUM BOXES
(5–6 in by 7–8 in/12–15 by 17–20 cm):
WRAPPING PAPER: 1 sheet, 15 by 15 in/38 by 38 cm
RIBBON: 24 in/61 cm

LARGE BOXES
(7–8 by 9–10 in/17–20 by 23–25 cm):
WRAPPING PAPER: 1 sheet, 20 by 20 in/50 by 50 cm
RIBBON: 30 in/76 cm

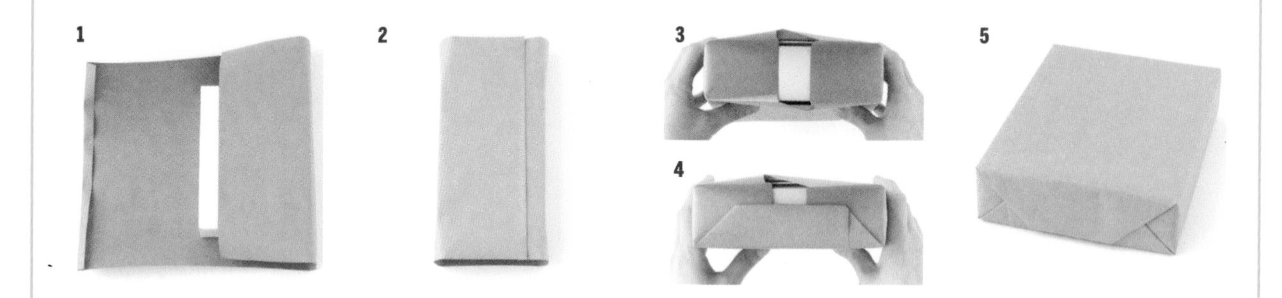

GIFT WRAP OPTIONS

BASIC GIFT WRAP 1

Ideal for tall, square boxes

1 Place the box in the center of the gift wrap. Fold the left edge over ½ in/12 mm. Fold the right flap over the gift and secure to the underside of the gift with double-sided tape.

2 Fold the left flap over and secure with double-sided tape.

3 On one end of the box, fold in both side flaps.

4 Fold the bottom flap up and secure with double-sided tape.

5 Fold the top edge under ¼ in/6 mm, fold the top flap down over the end of the gift, and secure with double-sided tape.

6 Repeat steps 3 through 5 on the other end of the box.

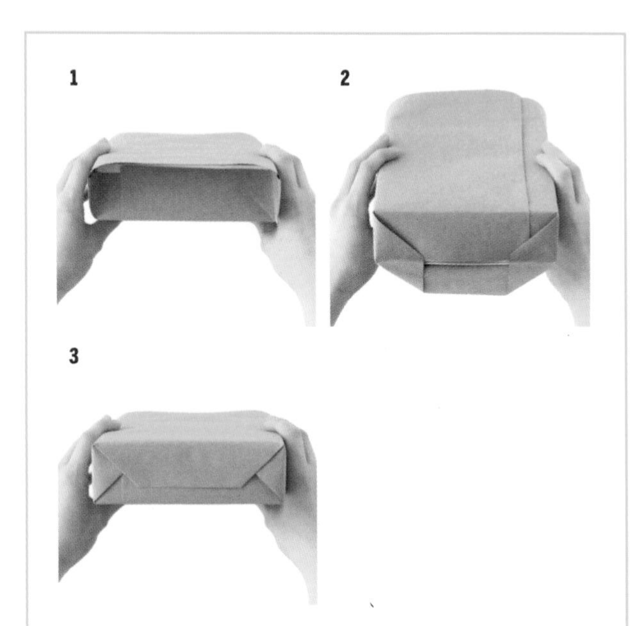

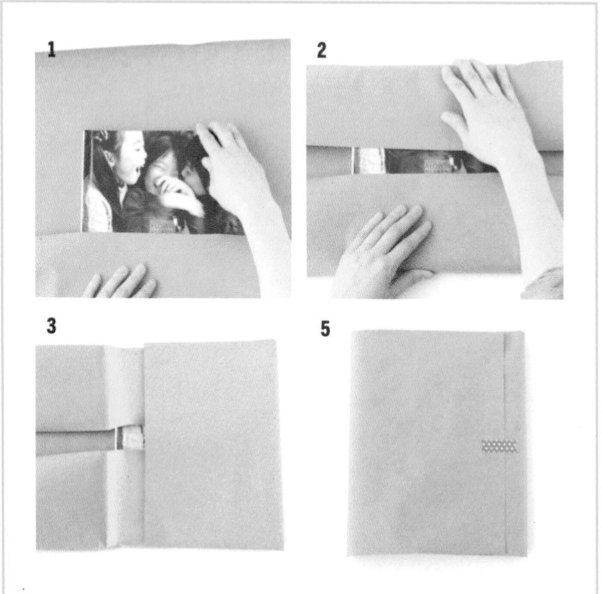

BASIC GIFT WRAP 2

Ideal for short, rectangular boxes

1. Follow steps 1 and 2 of Basic Gift Wrap 1 (facing page).
2. Fold the bottom flap up and secure with double-sided tape.
3. Fold in both sides.
4. Fold the top edge under ¼ in/6 mm, and fold the top flap down and secure with double-sided tape.
5. Repeat steps 2 through 4 on the other end of the box.

BOOK WRAP

1. Cut a piece of paper twice the width of the book. Center the book in the middle of the paper and allow for an overhang of half the length of the book on each end. Fold the bottom flap up and crease at the fold. For precise creases, use a bone folder.
2. Fold the top flap down and crease at the fold.
3. Put double-sided tape along the edge of the right flap and fold it in toward the book. Crease at the fold.
4. Fold the left edge under ½ in/12 mm. Put double-sided tape along the edge of the left flap and fold it in toward the book. Crease at the fold.
5. **OPTIONAL:** Add a piece of fabric tape (use glue stick on the back of the fabric piece).

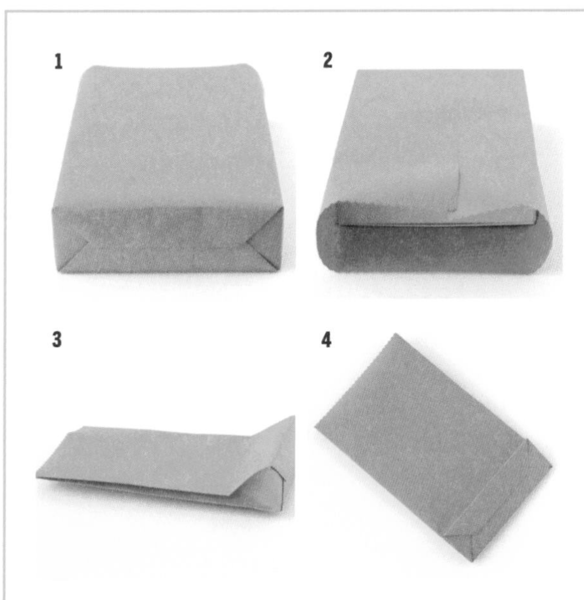

CUSTOM PAPER GIFT BAG

Create a custom gift bag in any size using boxes you have around the house (cereal boxes, shoe boxes) as templates. You can decorate these simple paper gift bags using many of the project techniques, including painting with watercolor, embellishing with stickers, or sewing on details. For precise creases, use a bone folder.

1 If you want a decorated edge on your bag, use decorative scissors to cut one long side of the paper. Then, follow steps 1 through 5 of Basic Gift Wrap 1 (page 12) on one side of the box to make the bottom of the gift bag.

2 Slide the box out of the bag. Secure the inside seams with tape.

3 Pinch in each side of the bag to create a crease in the middle.

4 Fold the bottom over so the paper bag can lie flat.

5 OPTIONAL: Close the bag with a piece of washi tape. Or you can punch two holes at the top of each side of the bag and make handles using string, ribbon, or twine.

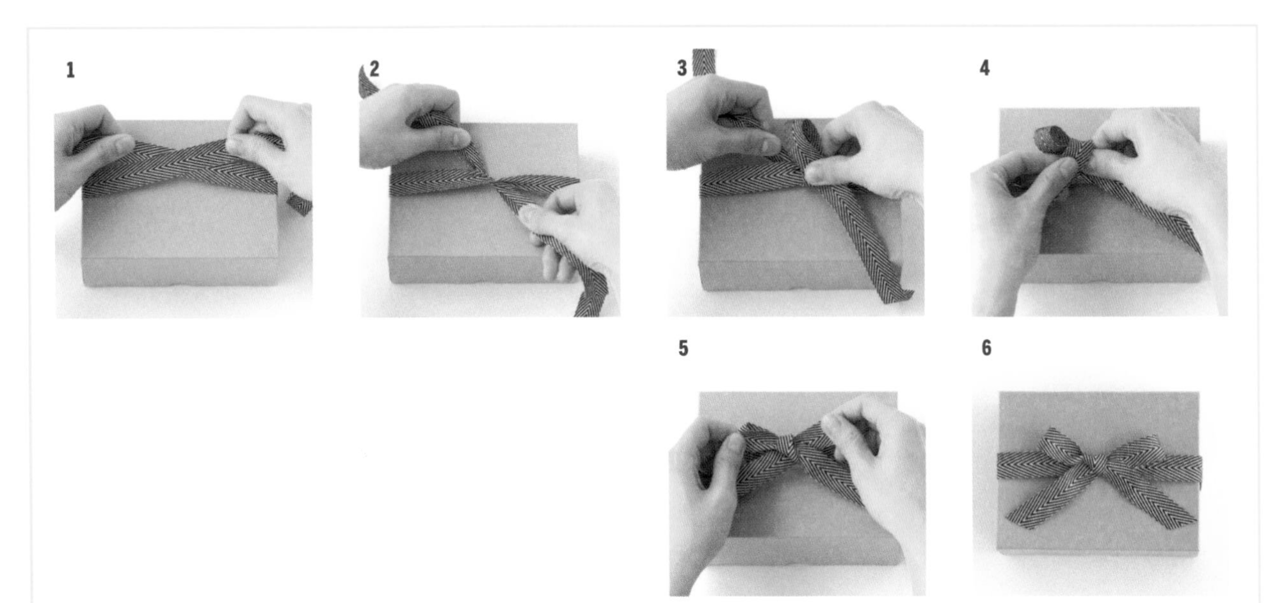

RIBBON TYING OPTIONS

You can use ribbon to add a pretty embellishment or a pop of color to your gift. Here are some basic methods of tying.

SIMPLE BOW

1 Use a length of ribbon that can be wrapped around the box twice. Place the ribbon on a flat surface and put the package in the middle of the ribbon. Wrap the two ends of the ribbon over the top of the gift. Hold the ribbon tightly around the box.

2 Use your right hand to pull the ribbon on the left side over and under the other end of the ribbon, creating a tie.

3 Put your right-hand thumb on the tie to hold it in place. Create a loop with the ribbon on the right side.

4 Move the loop to the left side and wrap the ribbon from the left side around and over the loop.

5 Pull the ribbon under and create a second loop over the first loop. Pull and tighten the loops to create a symmetric bow. If the ribbon is not snug enough around the package, untie it and try again.

6 Place the two ribbon tails together and cut them at an angle at the same place.

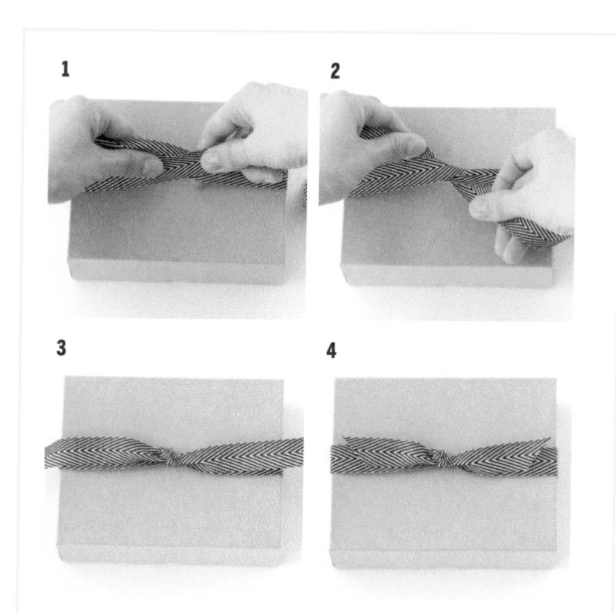

SIMPLE KNOT

1 Use a length of ribbon that can be wrapped around the box one and a half times. Place the ribbon on a flat surface and put the package in the middle. Wrap the two ends of the ribbon over the top of the gift. Hold the ribbon tight around the box.

2 Use your right hand to pull the ribbon on the left side over and under the other end of the ribbon, creating a tie.

3 Put a finger from your left hand on the tie to hold it in place. Use the ribbon ends to tie a knot. Pull the ribbon ends tight horizontally. Straighten the ribbon ends and flatten the knot.

4 Trim the ends of the ribbon by putting the ends together and cutting them at the same place at the same angle.

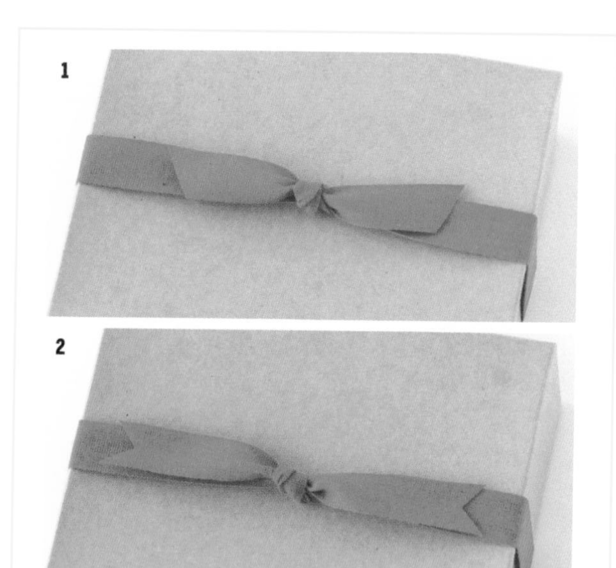

CUTTING RIBBON ENDS

After tying the ribbon in a simple bow or knot, cut the ends with a pair of sharp scissors for a clean finish. Here are two options for cutting the ribbon ends.

1 DIAGONAL CUT: Hold the two ribbon ends up together and cut at the same diagonal angle.

2 INVERTED V CUT: Fold one ribbon end in half and clip an inverted V from the folded part to the edge of the ribbon. Repeat on the other ribbon end.

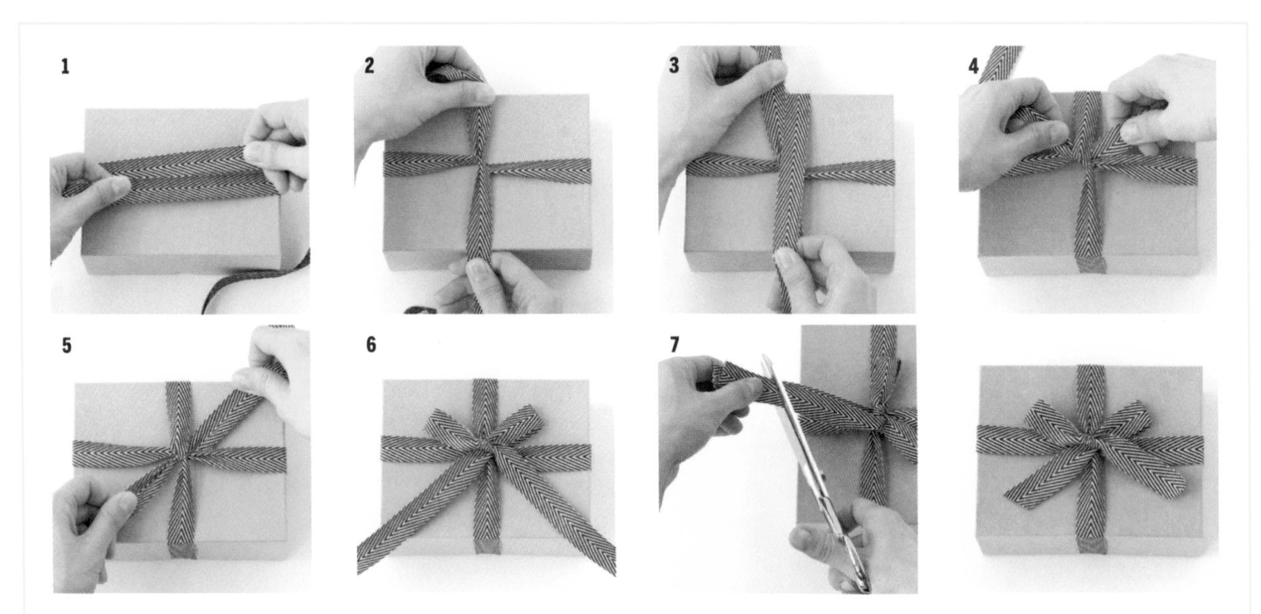

CROSS-TYING

1 Use a length of ribbon that can be wrapped around the box twice. Put the box on top of the ribbon. The length of the ribbon to the right of the box should be one-quarter of the length of the entire ribbon. Wrap the left section of the ribbon over the right section.

2 Cross the ribbon at the center of the box, holding the ribbon vertically.

3 Wrap the long ribbon end around the bottom and over the top of the box.

4 Place the long ribbon end through the bottom left corner of the center twist and up through the right corner.

5 Pull the ribbon ends toward the opposite corners.

6 Tie a simple bow using the instructions on page 15.

7 Trim the ends of the ribbon with a diagonal cut (see facing page).

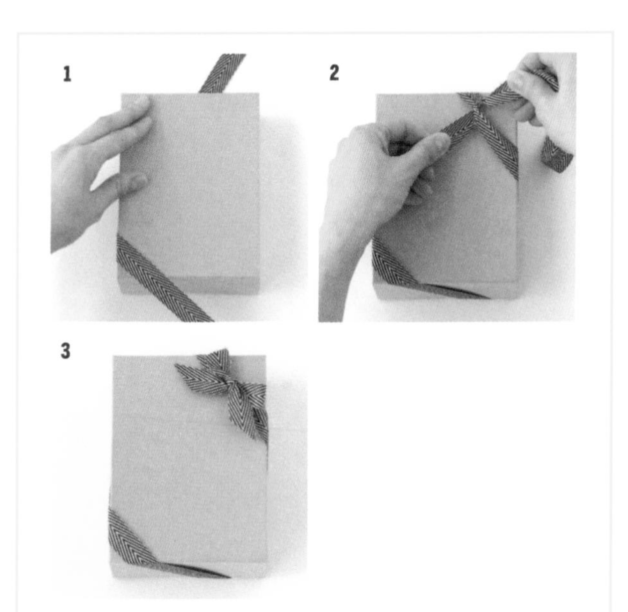

DIAGONAL TYING

1 Use a length of ribbon that can be wrapped around the box twice. Put the box on a flat surface. Place the ribbon underneath the box at a diagonal with one-quarter of the ribbon hanging at the top of the box.

2 Place the ribbon diagonally across the bottom left corner and back around to the top corner.

3 Tie a knot at the top corner and make a simple bow using the instructions on page 15. Adjust the bow to be centered on the corner of the box.

STRING TYING OPTIONS

1 STANDARD CROSS-PATTERN: Use a length of string that can be wrapped around the box twice. Hold the string vertically across the middle of the box, flip the box over, and cross the string at the center of the box, pulling the string horizontally. Flip the box over and tie the string in a bow.

2 MULTIWRAP: Wrap the string multiple times around the box and cut, leaving two short tails. Tie the two tails in a knot, and trim the ends.

3 BOX WITH STRING LACING: Use a leather hole punch and mallet to make two small holes in the top of a box. Thread string through the holes and tie on a gift tag. Secure the string with a knot on the inside of the box top.

(FIG. 1)

(FIG. 2)

(FIG. 3)

BONE FOLDER TECHNIQUES

SCORING (FIG. 1)

Scoring creates an indented line on the paper, which helps you to fold the paper crisply.

Starting from the point farthest away from you, hold the bone folder tip perpendicular to the paper and press the tip firmly on the paper, pulling it down toward you. Use the ruler as a guide to create a straight score on the paper.

FOLDING (FIG. 2)

A bone folder is also used to help create a neat crease when folding paper. This is particularly useful when working with card stock or heavyweight paper.

Once you have made the fold in your paper, press the bone folder firmly against the crease and move it along the folded edge.

ATTACHING TWO PIECES OF PAPER TOGETHER (FIG. 3)

A bone folder can also be used when adhering two pieces of paper together.

To adhere the papers after adhesive has been applied between them, run the long side of the bone folder over the surface of the paper.

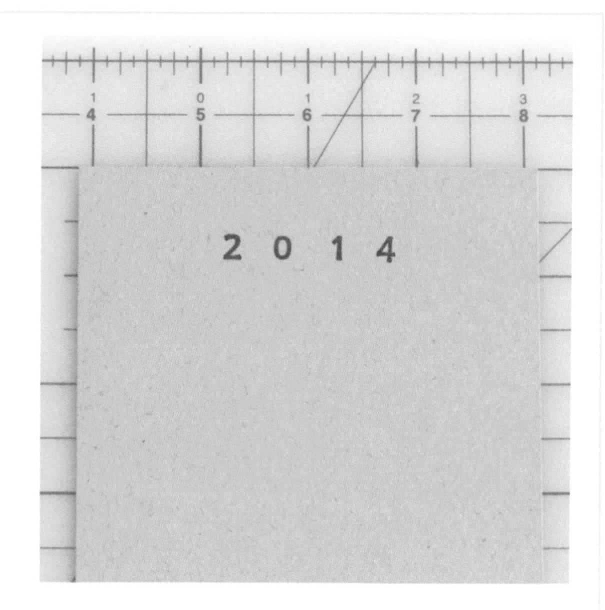

WRITING NAMES AND MESSAGES

When adding text to a gift tag or box, you have several choices: hand letter the recipient's name or a special message using a pen (gel, metallic, paint) or chalk pencil, create simple gift tags on the computer using fonts from various online font shops (many are free), use letter stickers, or use rolling or individual stamps to spell out a message. If you need help aligning alphabet or number stamps, see facing page for some tips.

INKING RUBBER STAMPS

Archival inks are the best for using on paper projects. Be sure to leave adequate time after stamping to allow the ink to dry (I recommend 10 minutes). To ensure that there is an even layer of ink on your rubber stamp so that you will produce clean and clear stamps, it helps to hold the rubber stamp face side up and place the ink pad on top of the rubber stamp.

1

2

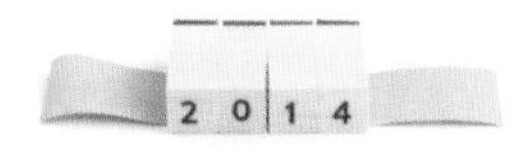

3

4

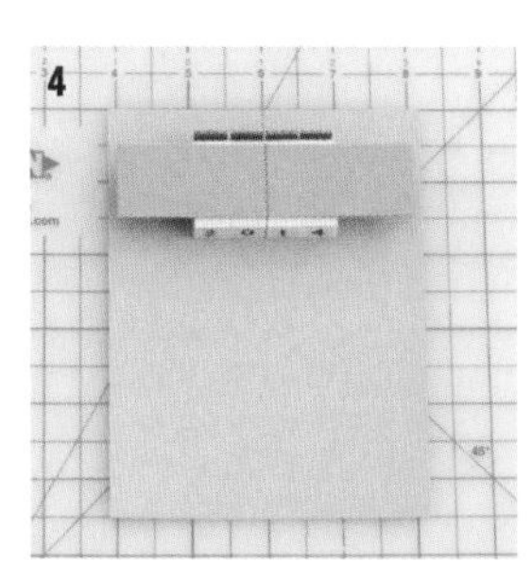

5

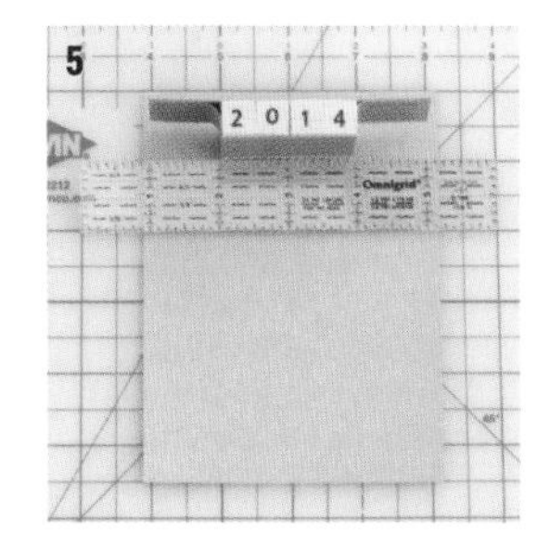

HOW TO CENTER ALPHABET or NUMBER STAMPS

MATERIALS

1 piece of cardboard 1 by 4 in/2.5 by 10 cm (length will vary depending on number of rubber stamps used)

Number or alphabet rubber stamp set

1 piece of adhesive artist tape, 4 in/10 cm

Piece of paper

Archival stamp pad

TOOLS

Black pen

Cutting mat

Pencil

Ruler

1 On one side of your cardboard, use a black pen to draw a line in the center. On the opposite side, draw a center line and parallel lines every ½ in/ 12 mm (the width of one stamp).

2 Put a piece of artist tape, sticky side up, on a flat surface. Put the desired rubber stamps right-side up on the artist tape, making sure they are straight.

3 Put the cardboard piece on top of the rubber stamps, aligning the center of the rubber stamps to the center marked on the cardboard. Secure the ends of the artist tape to the cardboard.

4 Place the paper that will be stamped on a cutting mat. Use a pencil to mark the center of the paper you want to stamp. Align the center mark on the cardboard with the center mark on the paper.

5 Using a ruler to keep the stamps straight, ink and press the rubber stamps onto the paper.

PROJECTS **PART 1**

WRAPPING PAPER

I have always enjoyed wrapping gifts with paper. Whether it's wrapping up a birthday present for my sons' friends or Christmas gifts for family, I love picking out paper to wrap gifts. In this section you will learn how to make your own pretty wrapping paper using basic supplies.

1 Hand-Carved Stamped Wrapping Paper
2 Photograph Wrapping Paper
3 Stitched Wrapping Paper
4 Patterned Wrapping Paper
5 Graph Paper Gift Wrap
6 Dotty Wood Wrapping Paper
7 Handmade Glitter Tape Gift Wrap
8 Vellum Stickers Wrapping Paper
9 Label Stickers Gift Wrap
10 Letter Stickers Gift Wrap
11 Polka-Dot Stickers Gift Wrap
12 Painted Dots Gift Wrap
13 Watercolor Wrapping Paper
14 Hand-Painted Furoshiki

HAND-CARVED STAMPED WRAPPING PAPER

Hand-carving stamps is a fun and simple craft that adds a special handmade detail to your gift. There are many beautiful wrapping paper prints out there, but I love the idea of printing your own. If you don't have time to carve your own stamp, you can always use a store-bought stamp or have a custom stamp made for you at your local stationery store.

MATERIALS

1 piece of scrap paper
Carving block
Archival stamp pad
Paper for wrapping your gift

TOOLS

No. 2 pencil
Bone folder
Speedball Linoleum Cutter Handle
Speedball Linoleum Cutter Blades Nos. 1 and 5
X-ACTO knife

HOW TO

1. Using the No. 2 pencil, draw your stamp design on the scrap paper.
2. Flip the paper right-side down onto the carving block and rub the back of the paper using the bone folder.

continued . . .

Try using several different colored stamp pads.

You can use a stamp to decorate stationery and paper gift bags or even pair your stamp with a fabric ink stamp pad to create custom drawstring bags and furoshiki cloth wraps.

3 Lift off the paper and you will see the transferred design on the carving block. Using a No. 1 linoleum carving cutter, carve out the outline of the stamp design. If the design has details, carve the negative space from the design.

4 When you are finished carving the stamp design, use the No. 5 carving cutter to carve a thick outline around the design. This will help you cut the stamp from the block.

5 Hold the block with one hand and use the X-ACTO knife to carefully cut the stamp image from the block. Please cut with care, and avoid cutting your fingers.

6 Hold the stamp right-side up and press the stamp pad onto the stamp, applying an even layer of ink. Test the stamp on the scrap paper to make sure it prints evenly. If there are any areas that do not stamp clearly, go back and use the No. 1 linoleum cutter to make the stamp edges more crisp. Once you have a stamp that produces a clear image, place the wrapping paper on a flat surface, right-side up, and stamp a pattern. Make sure to apply even pressure on the stamp to ensure a clean printed image. You can stamp a random or repeated pattern.

7 Let the stamped wrapping paper dry for 1 hour before wrapping your gift.

8 Wrap up your gift!

PHOTOGRAPH WRAPPING PAPER

Photos aren't just for hanging on walls or placing inside albums. You can turn your favorite photo into wrapping paper by printing it on your home printer, or you can take it to your local print shop and have it enlarged into an engineer-size print, suitable for wrapping oversize boxes. You can have the photographs printed in color, or in black and white for a classic look.

MATERIALS

Digital photograph

Text-weight paper

TOOLS

Computer and printer

1. Print the photograph onto the text-weight paper.
2. Center the main focus of the photograph on the center of your gift.
3. Wrap up your gift!

You can scan a child's artwork, print it out, and turn it into wrapping paper.

You can also create a collage of photographs on the computer, print it out, and turn it into wrapping paper.

STITCHED WRAPPING PAPER

MATERIALS

Cotton thread in two colors
Butcher paper

TOOLS

Sewing machine or hand-sewing needle
Pencil (if sketching a design)
Scissors

This stitched wrapping paper gives a fresh look to white butcher paper. People often think you only use a sewing machine or needle and thread to work with fabric, but there are many fun and interesting ways to use thread with paper to add pretty details and texture. You can lightly draw a design with a pencil and stitch over the design or create your own freestyle design. Even if you don't have a sewing machine, you can easily achieve a similar look by hand-sewing with a needle and thread.

BRIGHT IDEAS

If your sewing machine has different stitch options, try using a zigzag or wavy stitch.

To add visual interest, try folding pleats in the paper and stitch along the edge of each pleat to secure it.

HOW TO

1. Thread your sewing machine (or needle) with cotton thread. Sketch a design onto your paper with a pencil, if desired. The center of the wrapping paper will end up being on the top of the wrapped gift box, so make sure there is a sufficient amount of design in the center area. Depending on the size of your paper, you might have to roll the paper under itself in order for it to fit in the sewing machine.

2. Sew lines that follow your sketch, or sew random lines, to create a stitched pattern on the butcher paper.

3. Thread your sewing machine (or needle) with the second color cotton thread. Repeat step 2.

4. Use scissors to trim excess thread from the edges of the wrapping paper.

5. Wrap up your gift!

PATTERNED WRAPPING PAPER

MATERIALS

Butcher paper

Paint pens, metallic pens, or marker pens

TOOLS

Extra piece of paper to protect work surface

Masking tape (optional)

Butcher paper makes a wonderful base for creating your own personalized decorative gift wrap and is a simple, inexpensive alternative to store-bought wrapping paper. You can find butcher paper in rolls in an array of colors at office or party supply stores. You can use paint pens, metallic pens, or markers to decorate the paper with fanciful patterns: draw polka dots, stripes, flowers, or geometric shapes. Allow time for the paints or inks to dry before wrapping gifts.

1. Place the extra piece of paper underneath the butcher paper to protect your work surface from pen marks. If using, place a small piece of masking tape on each corner of the butcher paper to keep it flat.
2. Starting from one side of the butcher paper, use the pens to draw a pattern on the paper. Let the ink dry for 1 hour.
3. Wrap up your gift!

BRIGHT IDEAS

Try making a gift bag from your patterned paper using the instructions on page 14.

Use patterned paper as a contrasting decorative band on a gift by cutting a strip of the paper, wrapping it around the middle of the wrapped gift, and securing it underneath with double-sided tape.

GRAPH PAPER GIFT WRAP

MATERIALS

Graph paper

Fine-tip pen

One of my favorite things to shop for in office supply stores is graph paper. If you do a bit of looking around, you can find graph paper with graph lines in a variety of colors, such as blue, green, gray, and black. Graph paper vellum, which is usually translucent, is fun to layer over a solid-color wrapping paper. And the uniform squares in the graph paper make perfect grids for writing special messages to the recipient.

BRIGHT IDEAS

Use a date or time stamp on the graph paper to add a meaningful date or time for the recipient.

Using a colored marker or pencil, color in alternating squares to make a checkered pattern.

1. Wrap up your gift with the graph paper.
2. With the fine-tip pen, write a message to the recipient inside the squares of the graph paper.

DOTTY WOOD WRAPPING PAPER

I love the organic look of wood veneer. Although we might not typically associate wood veneer with gift wrapping, it can provide a natural element to your gift. When I discovered these wood dot stickers at a woodworking store, I immediately scooped up several packs. You can use them to make a whimsical dotty pattern on wrapping paper.

MATERIALS

Wood veneer dot stickers
Wrapped gift
Pencil (optional)

BRIGHT IDEAS

Use a fine metallic paint pen to write the initials of the recipient on the wood dot stickers.

Create patterned dots. Apply masking tape to cover half of each dot sticker. Apply two layers of acrylic paint to the exposed halves of the dot stickers and let dry for 3 hours. Remove the tape from the stickers and you'll have two-toned stickers.

Place the wood veneer dot stickers on your wrapped gift to create a vertical wave pattern. You can also use a pencil to draw a pattern and then place the stickers on top of the drawn pattern.

HANDMADE GLITTER TAPE GIFT WRAP

MATERIALS

Dressmaker's tape measure

Wrapped gift

½-in-/12-mm-wide double-sided tape with backing

Fine glitter

TOOLS

Scissors

2 pieces of paper

Small brush (optional)

There's nothing like a little sprinkle of glitter to add a magical touch to your gift. Making glitter tape is easy and there are many options for applying it to your gift. Touches of glitter look best against solid-colored wrapping paper. Since a dressmaker's tape measure is flexible, it comes in handy when measuring the length around a box.

1 Wrap the dressmaker's tape measure around your wrapped gift to measure the distance around the gift, or twice its width plus twice its depth.

2 Use scissors to cut a strip of double-sided tape 1 in/2.5 cm longer than the total measurement obtained in step 1.

Cut the glitter tape into shapes and add them to the top of your wrapped gift.

You can use patterned washi tape instead of making glitter tape.

3. Place the tape sticky-side up onto one piece of the paper. If the ends of the tape curl up, place a small piece of the tape on each end to secure it to the paper. Sprinkle the glitter on the sticky tape so the entire surface is covered. Remove the tape that is holding down the ends of the sticky tape, if using. Shake off the excess glitter onto the second piece of paper. Use the paper to funnel the excess glitter back into the glitter dispenser.

4. Remove the backing from the double-sided tape and place it on your wrapped gift, overlapping the ends on the underside of the gift.

   NOTE: Use a small brush to wipe off excess glitter that may fall onto the wrapping paper, or use a small piece of tape to pick it up.

VELLUM STICKERS WRAPPING PAPER

MATERIALS

Vellum paper in a variety of colors

Wrapped gift

TOOLS

Circle paper punches in various sizes

Sticker machine

Stickers can transform a plain package into a delightful gift. Two of my favorite crafting tools are a sticker maker and a paper punch. You can transform any paper shape into a sticker by using a sticker machine, which adds adhesive to the back of any paper that will fit through the opening of the machine. You can also turn ribbon or fabric into stickers. If you don't have access to a sticker maker, you can use a glue stick to adhere the vellum paper circles to the gift.

1. Using the circle paper punches, cut out rounds of different sizes from the vellum paper.
2. Place the paper rounds in the sticker machine, and make the stickers. Remove the sticker backing.
3. Put the stickers on the wrapped gift in a pleasing pattern.

You can use colored text-weight paper or cardstock instead of vellum paper for a bold look.

Cut out faces in photographs and make them into stickers.

LABEL STICKERS GIFT WRAP

MATERIALS

Wrapped gift

TOOLS

Label maker

Scissors

Trim the ends of the labels to make them into a banner sticker.

Use labels of different colors to create a colorful message on your gift.

Label makers help keep the house and office organized and are also handy when you want to personalize a gift. Label refills come in a variety of colors and fonts. With these labels, you can write out a name or create a special message to the recipient.

1. Using the label maker, create a special message for your recipient. Use scissors to cut out the label and remove the backing.

2. Apply the label to your wrapped gift.

LETTER STICKERS GIFT WRAP

Letter stickers are another easy way to personalize a package. You can find a wide assortment of letter stickers at your local office supply store or craft store. Letter stickers come in matte colors or metallics and glitters.

MATERIALS

Wrapped gift

Letter stickers

On the top surface of the wrapped gift, use the letter stickers to write out the recipient's name or a special message to the recipient.

You can make your own letter stickers by using a letter stencil on contact paper. Cut the letters out with scissors.

Use colored masking tape to create letters.

POLKA-DOT STICKERS GIFT WRAP

MATERIALS

Text-weight colored paper
Wrapped gift
Double-sided tape
Coding color labels

Create a pattern or write out the initials or name of the recipient with stickers for a personalized touch. You can find a wide selection of stickers at office supply stores. My favorite stickers are the office coding color labels, which come in a beautiful array of colors and sizes.

Add washi tape stripes to give it a different look.

Write out a message by writing individual letters in the stickers.

1. Wrap the text-weight colored paper around the wrapped gift and secure with the double-sided tape.
2. Place the coding color labels on the gift to create the first and last initials of the recipient's name.

PAINTED DOTS GIFT WRAP

Hand-printed patterns on wrapping paper add personalized charm. Creating your own polka-dot pattern is as simple as using acrylic paint and a pencil eraser as a stamp. You can use several different paint colors or stick to one color, creating a random or systematic pattern.

MATERIALS

Colored paper

Acrylic paint

TOOLS

Extra piece of paper to protect work surface

Small paper plate

Pencil with eraser

1. Lay out the extra piece of paper on a flat surface, and place the colored paper on top. Place a dab of the paint on the paper plate. Dip the pencil eraser into the paint and blot it on the side of the paper plate to remove excess paint. Create a pattern of dots all over the paper, repeatedly dipping the pencil eraser into the paint to re-ink.
2. Let the paint dry for 3 hours.
3. Wrap up your gift!

BRIGHT IDEAS

For large dots, use round sponges.

You can also use a stamp and stamp pad to create a pattern on tissue paper.

WATERCOLOR WRAPPING PAPER

Watercolor is a fun and versatile way to create lovely patterned wrapping paper. You can paint abstract or floral patterns using several different colors or you can make an ombré design by painting stripes in progressively darker shades of your favorite color. Try experimenting with different paintbrush sizes for varied effects.

MATERIALS

White paper

Watercolor paints in a variety of colors

Water

TOOLS

Extra piece of paper to protect work surface

Paint mixing tray

Paintbrush

2 containers for water (1 for mixing colors, 1 for cleaning the brush)

Paper towel

BRIGHT IDEAS

Instead of using a paintbrush, dilute the watercolor paint with water, pour it into a spray bottle, and spray the paint onto the paper for a Jackson Pollock effect.

If you want to minimize the puckering of the damp paper, place a layer of paper towels over the paper while it's drying and then put heavy books on top of the paper towels.

HOW TO

1 Place the extra piece of paper underneath the white paper you will be painting, to protect your work surface. Place a dot of paint on the mixing tray and add a little water. Mix the paint and water with the brush until the desired shade is achieved.

2 Dip the paintbrush into the paint and make strokes across the paper in an abstract pattern. When changing to another paint color, first rinse the paintbrush in water and pat dry on the paper towel. Continue painting until the desired pattern is achieved on the paper. Let the paint dry for 3 hours.

3 Wrap up your gift!

HAND-PAINTED FUROSHIKI

A furoshiki is a traditional Japanese cloth wrapper for gifts, which is not only beautiful but also eco-friendly and reusable. Recipients will keep the gift wrap for years to come. You can find many lovely fabric prints, or you can make a unique furoshiki by painting a pattern onto cotton fabric. Make sure to choose a lightweight cotton fabric for easy folding. I've specified the dimensions for a traditional furoshiki here, but you can choose a larger or smaller piece of fabric, depending on the size of your box—just be sure the fabric is a square.

MAKES

1 furoshiki, 28 by 28 in/71 by 71 cm

MATERIALS

- 1 piece of lightweight cotton fabric, 29 by 29 in/73.5 by 73.5 cm
- Fabric paint
- Cotton thread
- Gift box

TOOLS

- 1 piece of cardboard, 30 by 30 in/76 by 76 cm
- Masking tape
- Shallow plastic bowl
- Foam paintbrush
- Iron
- Ironing board
- Ruler
- Scissors
- Straight pins
- Sewing machine (or hand-sewing needle)

BRIGHT IDEAS

Use a hand-carved stamp (see page 25) or store-bought stamp with a fabric ink stamp pad on plain cotton fabric to create a custom furoshiki.

Instead of painting your own furoshiki, use vintage scarves, tea towels, or store-bought printed fabric (rectangle fabric pieces will need to be cut down and sewn into a hemmed, square piece).

continued . . .

1. Place the cardboard on a flat surface. Lay the cotton fabric on top of the cardboard, right-side up. Place a strip of masking tape diagonally on the fabric in one corner with the ends of the tape hanging ½ in/12 mm over both fabric edges. Press the tape down firmly on the fabric to prevent the fabric paint from bleeding under the tape. Repeat by placing strips of masking tape across the entire piece of fabric to the opposite corner. You can vary the width between the masking tape strips or keep it uniform.

2. Put a small amount of fabric paint in the plastic bowl. Using a sponge brush, apply the fabric paint on the fabric. Depending on the fabric and paint colors, you may need to apply more than one coat of paint. Let the fabric paint dry for 2 hours.

3. Remove the masking tape strips from the fabric by gently peeling back one end of the masking tape across the fabric. To heat-set the fabric paint, turn the iron on to medium-high heat with no steam. Place the fabric right-side down on the ironing board and iron the back of the fabric for 3 to 5 minutes. Make sure to apply even pressure over the entire fabric.

4. To create a flat crease on all four sides, keep the fabric right-side down, and on one side, use a ruler to fold the edges over ¼ in/6 mm to the wrong side and press with the iron. Moving clockwise, repeat on the remaining three sides. Fold each side ¼ in/6 mm to the wrong side a second time and press. Unfold the corners of the fabric and, with scissors, trim off the corner edges on a diagonal across the first fold lines.

5. Fold the new edge ¼ in/6 mm down to the corner of the second fold line and press with the iron.

6. Fold over the edges once, as you would when wrapping a box with paper, and press with the iron to form a mitered corner. Pin the mitered corner and fabric sides in place.

7. With the cotton thread and with the sewing machine or a hand-sewing needle, topstitch the folded edges ⅛ in/3 mm from the edge of the fabric.

8. To wrap your gift, lay the furoshiki on a flat surface, pattern-side down. Place the gift box diagonally in the middle of the square. Fold one corner of the cloth over the length of the box and tuck it under the box.

9. Repeat with the opposite corner of the cloth.

10. Working with the two remaining corners of the furoshiki, pinch the fabric in on both edges of the box.

11. Tie the two ends together in a double knot.

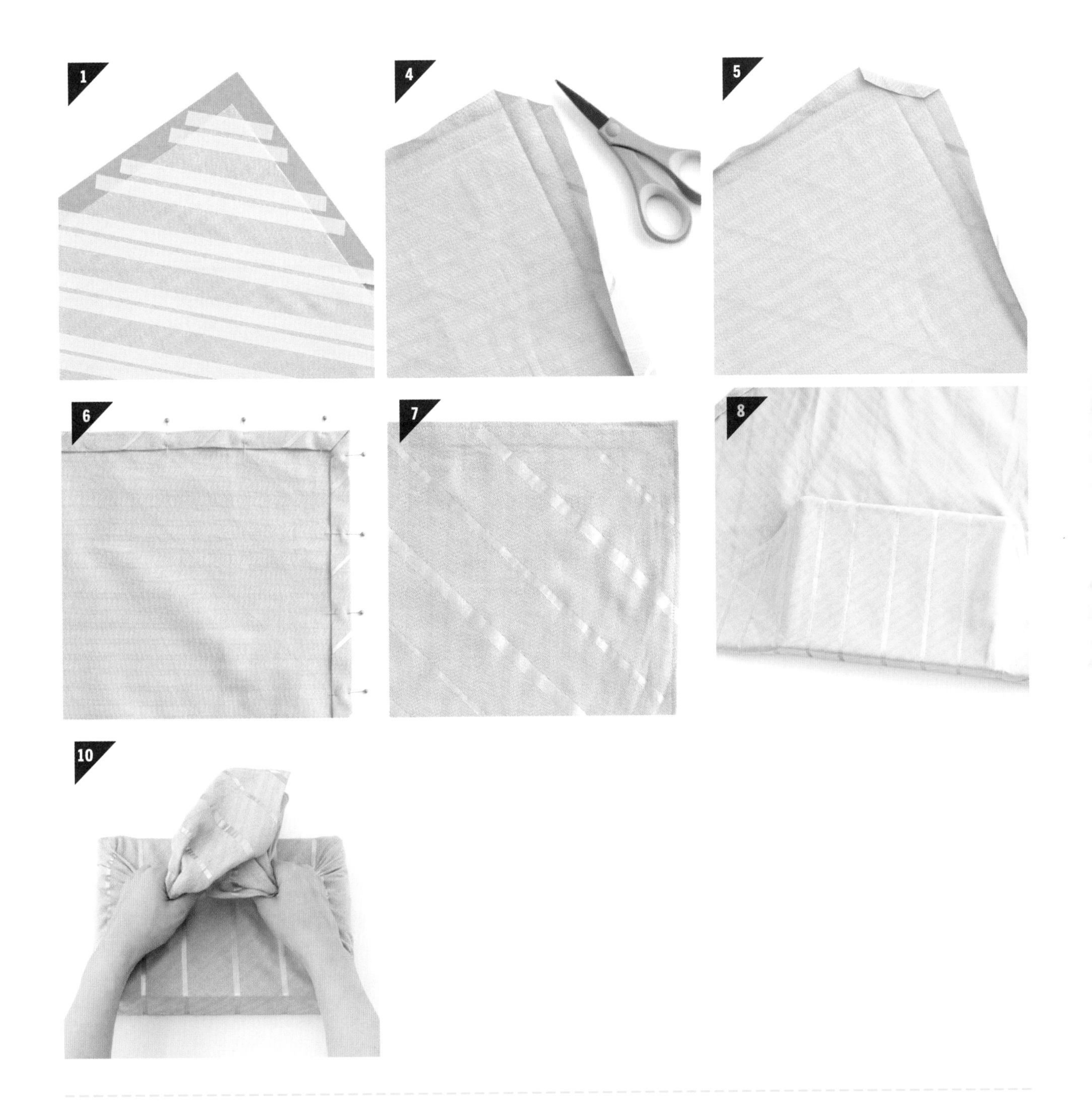
1
4
5
6
7
8
10

PROJECTS **PART 2**

GIFT BOXES, BAGS, AND TOPPERS

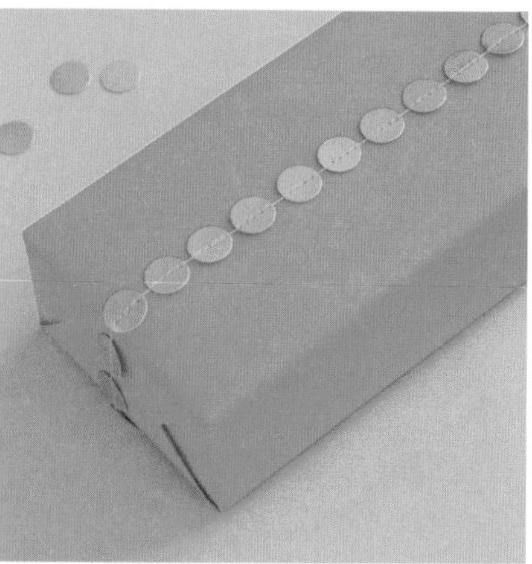

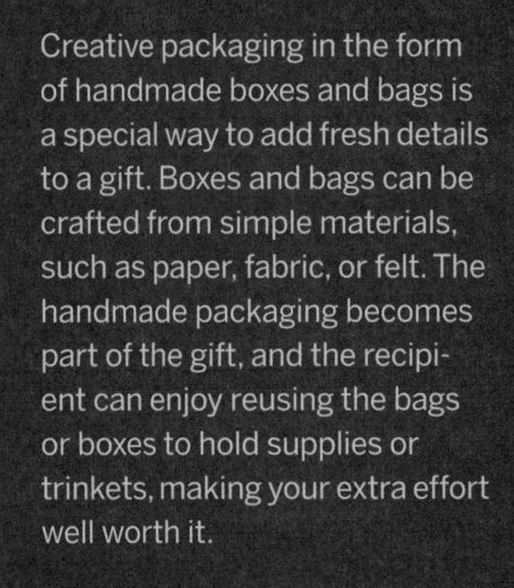

Creative packaging in the form of handmade boxes and bags is a special way to add fresh details to a gift. Boxes and bags can be crafted from simple materials, such as paper, fabric, or felt. The handmade packaging becomes part of the gift, and the recipient can enjoy reusing the bags or boxes to hold supplies or trinkets, making your extra effort well worth it.

15 Custom Cardstock Gift Box
16 Felt Gift Box
17 Washi Tape–Wrapped Tin
18 Custom Drawstring Bag
19 Personalized Fabric Bag
20 Washi Tape Glassine Bag
21 Stitched Felt Pouch
22 Yarn Pom-Pom Topper
23 Felt Letter Gift Topper
24 Fabric Flag Gift Topper
25 Wood Veneer Bow Gift Topper
26 Fabric Button Gift Topper
27 Tissue Paper Flower Gift Topper
28 Paper Rosette Gift Topper
29 Stitched Paper Garland
30 Felted Ball Garland
31 Brownie Mix Glass Jar Kit
32 Baked Goods Packaging

CUSTOM CARDSTOCK GIFT BOX

Once you know how to make your own gift boxes, you'll be able to create boxes in any color or pattern you like. Try using plain or patterned cardstock to make a custom box. You can use custom boxes to package gifts or party favors.

MAKES

1 box, 2 by 2 by 2 in/ 5 by 5 by 5 cm

MATERIALS

1 sheet of colored cardstock, 8½ by 11 in/21.5 by 28 cm

Transparent tape

TOOLS

Computer and printer

Custom Cardstock Gift Box Template

X-ACTO knife

Ruler

Cutting mat

Bone folder

1. Download the cardstock gift box template at www.chroniclebooks.com/prettypackages and print it out onto colored cardstock. Using the X-ACTO knife, ruler, and cutting mat, cut out the box.

2. With the cardstock still on the cutting mat, use the ruler and bone folder to score along the dashed lines.

3. Fold along the scored lines to create the box. Use the tape to close the seams.

Use patterned cardstock or scrapbook paper to make the box.

Use a stamp and stamp pad to create a pattern on the cardstock before making the box.

FELT GIFT BOX

A cozy felt gift box can hold a small gift, and then the gift recipient can use the felt box later to hold craft or office supplies. Felt comes in a beautiful array of colors and weights. Stiff felt is easy to work with because it keeps its shape and the edges do not fray when cut.

MAKES

1 box, 5⅛ by 5⅛ by 2 in/ 12.5 by 12.5 by 5 cm

MATERIALS

Two pieces of stiff wool felt; one 7 by 7 in/17 by 17 cm, one 5⅛ by 5⅛ in/12.5 by 12.5 cm

Hot glue stick

TOOLS

White chalk pencil

Ruler

Cutting mat

Bone folder

Scissors

Hot glue gun

1 Using the chalk pencil, ruler, and 7-by-7-in/17-by-17-cm piece of felt, draw vertical lines 2 in/5 cm and 5 in/12 cm from the left edge. Rotate the piece of felt 90 degrees and repeat.

BRIGHT IDEAS

Paint a pattern on the felt using fabric paint.

Make the lid and bottom of the box out of felts of different colors.

continued . . .

2 Using the chalk pencil, ruler, and 5⅛-by-5⅛-in/12.5-by-12.5-cm piece of felt, draw vertical lines 1 in/2.5 cm and 4⅛ in/10.5 cm from the left edge. Rotate 90 degrees and repeat.

3 Lay the felt pieces on the cutting mat. Using the bone folder, score along each white chalk line on both pieces of felt. This will help you fold the flaps in to create the box.

4 Using the scissors, cut along the two short vertical lines at the bottom of the larger felt piece and the two short vertical lines at the top of the larger felt piece. Repeat on the smaller piece of felt.

5 On one side of the larger felt piece, using the hot glue gun and hot glue stick, glue one short flap to an adjacent longer flap. Secure in place for a few seconds by holding the flaps together with your fingers. Repeat with the opposite long flap. Repeat on the other side of the box.

6 Repeat step 5 on the smaller piece of felt to make the lid.

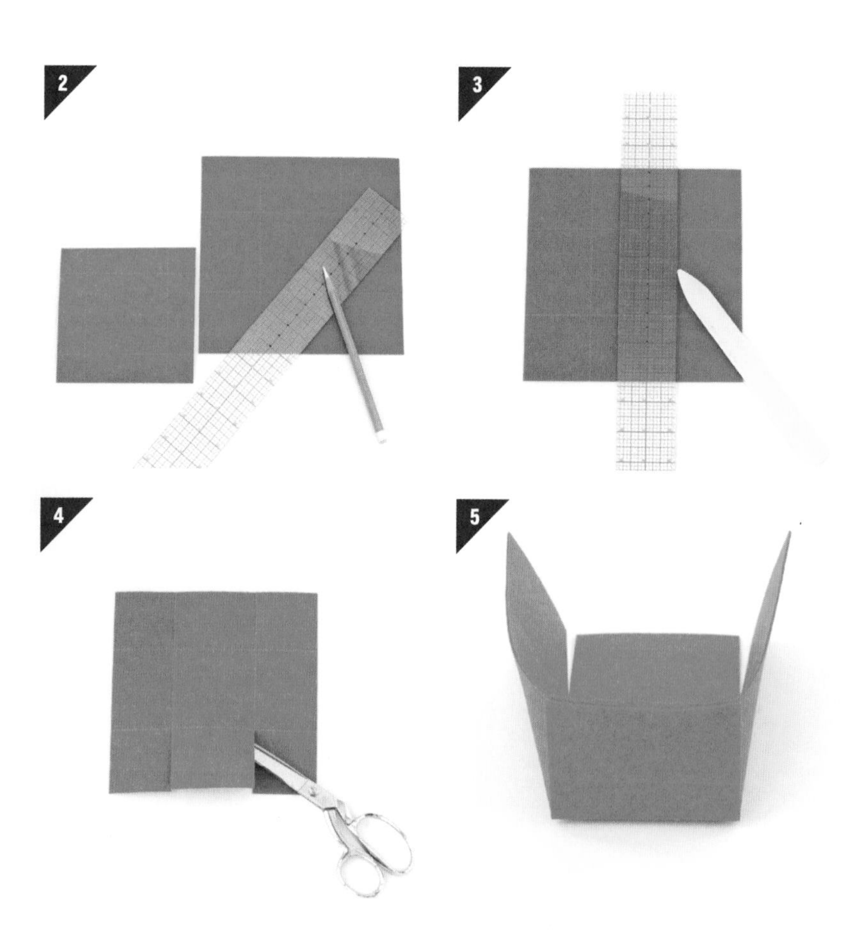

WASHI TAPE–WRAPPED TIN

This is a simple and quick project that can be made with items from around your home. Use paint cans or old tin cans, making sure to use a metal file to file down any sharp edges. I love that the tin becomes part of the gift and can be reused as pretty storage. Washi tape comes in a wonderful variety of colors and patterns, allowing you to customize the gift to suit the recipient's style.

MAKES

1 tin

MATERIALS

Tin can

Washi tape (in various colors or patterns)

TOOLS

Metal file (optional)

Scissors

If you are using an old tin can, use a metal file to file down any sharp edges. Start at the seam of the tin can and wrap the washi tape around the tin, snipping the tape with the scissors when it overlaps itself slightly. Vary the tape colors in whatever pattern you'd like.

BRIGHT IDEAS

Instead of stripes, try a plaid print by weaving the tape vertically and horizontally on the tin.

Try combining washi tape with colored masking tape, glitter tape, or artist's tape for a multitextured look.

CUSTOM DRAWSTRING BAG

Handmade drawstring bags are one of my favorite ways to package gifts. They require a bit of planning, but you can make a bag to wrap a gift of any size. I have made them as small as a deck of cards and as big as a laundry bag. You can select fabric in the recipient's favorite color or use a print that matches the theme of the present.

continued . . .

MAKES

One drawstring bag, 6½ by 7½ in/16.5 by 19 cm

MATERIALS

1 piece of fabric, 7 by 18 in/17 by 46 cm

Cotton thread

2 pieces of ribbon, 22 in/56 cm long and ½ in/12 mm wide

TOOLS

Straight pins

Sewing machine

Iron

Ironing board

Chopstick

Safety pin

Scissors

Substitute the ribbon with twill tape or trim.

Use a solid-color fabric and use fabric paint and a paint-brush to add a pattern.

1. Fold the fabric in half crosswise, right-sides together. Starting 2 in/5 cm from the top of the fabric, place straight pins along the two long sides of the fabric.

2. Thread your sewing machine with cotton thread. Starting 2 in/5 cm from the top of the fabric, sew down one side of the bag using a ¼-in/6-mm seam allowance, removing the straight pins as you sew. Repeat for the other side of the bag.

3. On one side of the bag, fold and iron a ¼-in/6-mm seam allowance on the right and left edges of the 2-in/5-cm portion that was not sewn. Fold over one more time and press with the iron. Repeat for the other side of the bag.

4. Turn the bag right-side out. With the seam centered, sew a ⅛-in/3-mm seam allowance along the unsewn V-shape at the opening on one side. Repeat for the other side.

5. Turn the bag wrong-side out. Fold under and iron a ¼-in/6-mm hem on the top two sides of the bag.

6. Fold the top side under again and iron, this time making a ¾-in/2-cm hem. Turn the bag over and repeat on the other side of the bag.

7. Turn the bag right-side out and use the chopstick to push the corners out. Sew across one side of the top of the bag, as close as you can to the lower edge of the hem. Repeat for the other side.

8. Clip the safety pin to the end of one piece of ribbon. Use the safety pin to thread the ribbon through the drawstring channel on one side of the bag and back out the other side, leaving both ends of that piece of ribbon hanging out on the same side of the bag. Repeat, starting with the drawstring channel on the opposite side. Clip the four ribbon ends to make them even. Tie the pairs of ribbon ends together.

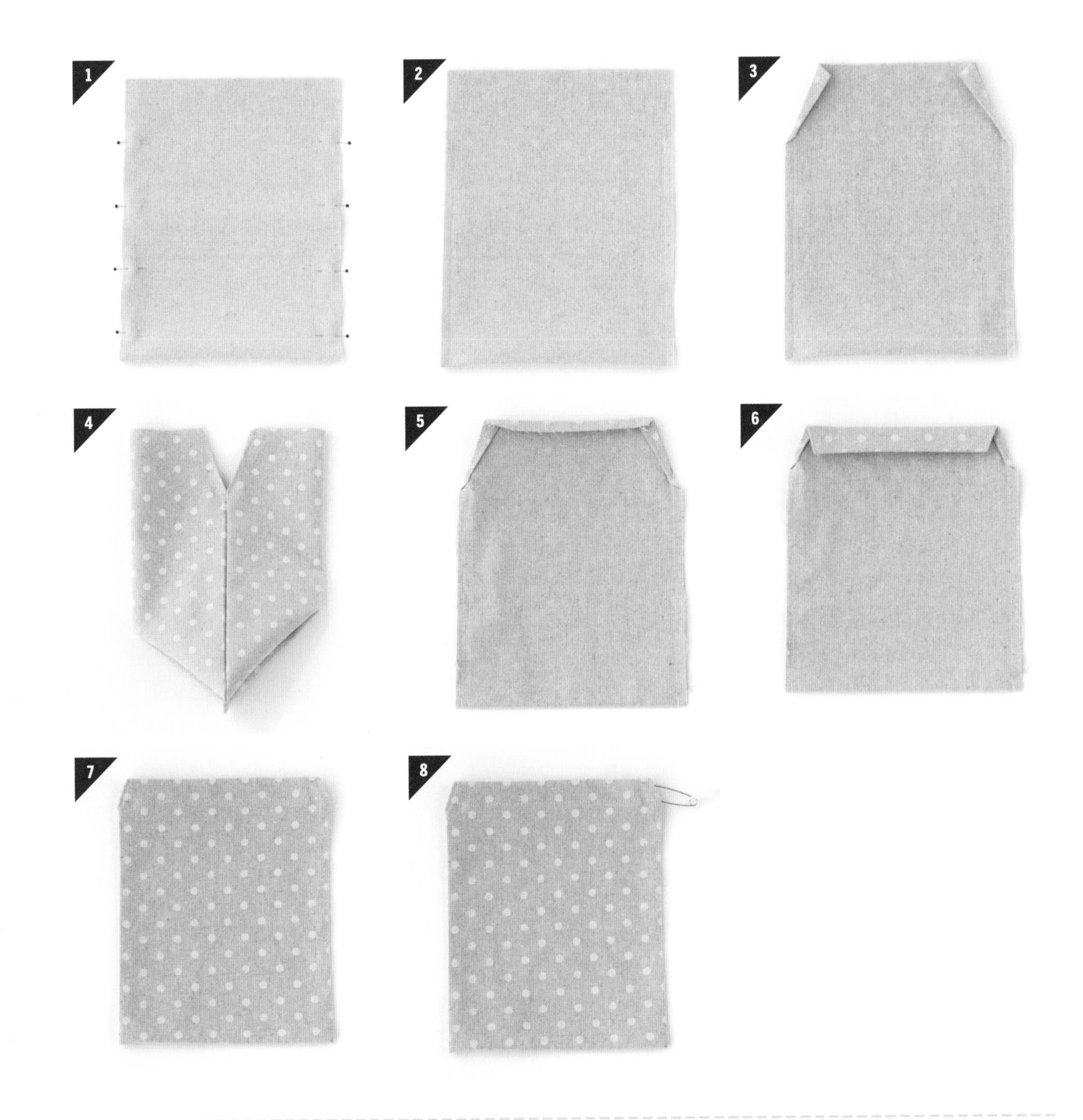
1
2
3
4
5
6
7
8

A

PERSONALIZED FABRIC BAG

I've been making personalized fabric bags for years. Kids and adults love receiving a personalized bag that they can use to hold their belongings. You can sew your own fabric bag using the instructions for the Custom Drawstring Bag (page 57), or you can go the quick route and use cotton or muslin bags sold in packaging supply stores online. Freezer paper has a matte side and a wax side. When you apply an iron to the matte side of the paper, the wax side temporarily adheres to the fabric—making it an ideal material for a stencil. Freezer paper stencils are fun to make and can be applied on bags, T-shirts, or anything made of fabric. You can find freezer paper at your local market.

continued . . .

MAKES

1 bag

MATERIALS

Muslin bag, 5 by 7 in/12 by 17 cm

Fabric paint

TOOLS

Pencil or printer

1 piece of freezer paper, 3½ by 4½ in/9 by 11 cm

Cutting mat

X-ACTO knife

Iron

Ironing board

1 piece of cardboard, 4 by 5 in/10 by 12 cm

Foam paintbrush

BRIGHT IDEAS

Use iron-on transfer letters found at fabric stores to personalize the bag.

Print your design onto iron-on transfer paper, cut it out, and iron it onto the fabric bag.

1. If creating your own bag, follow the instructions for making a drawstring bag on page 57.

2. Draw or print a letter onto the matte side of the freezer paper. Big, bold letters work best.

3. Put the freezer paper on the cutting mat and use the X-ACTO knife to cut out the letter. The portions of the letter that you cut out will be what is painted.

4. Using an iron on medium-high setting with no steam, iron the freezer paper stencil onto the muslin bag. If the letter stencil has a separate inside piece, carefully place it inside the stencil and iron onto the bag.

5. Place the piece of cardboard inside the muslin bag to prevent the paint from leaking to the other side of the bag. Dip the foam brush into the fabric paint and paint a layer over the letter shape. If you want more coverage, repeat with a second layer of paint. Let the paint dry for 6 hours. You can use a blow-dryer on high heat to speed up the drying process.

6. Carefully peel off the freezer paper stencil from the bag.

7. Place the bag face down on the ironing board, and with the iron set on medium heat, iron the bag from the back side to heat-set the fabric paint for 3 minutes.

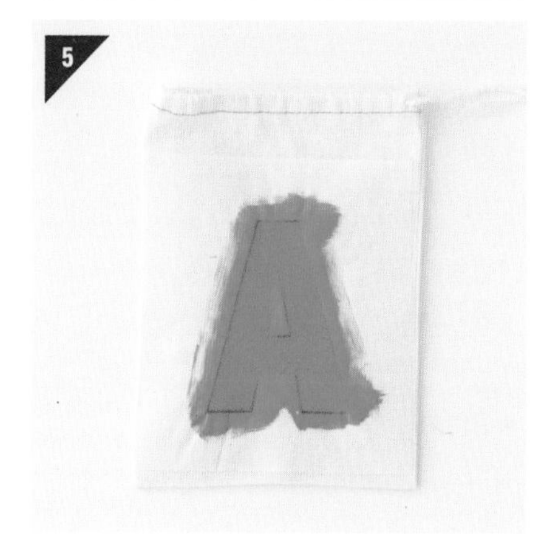

WASHI TAPE GLASSINE BAG

Glassine bags are transparent bags that are perfect for holding party favors, small gifts, or edible treats. They are ideal to use when you want to show what is inside. I keep a stash of glassine bags in various sizes in my studio and they always come in handy for packaging gifts. Washi tape comes in various colors, patterns, and widths, which you can also cut to your preferred width using an X-ACTO knife, ruler, and cutting mat.

continued . . .

MAKES

1 bag

MATERIALS

Washi tape in different colors and patterns

Glassine bag

TOOLS

Scalloped scissors (optional)

Cutting mat

X-ACTO knife

Ruler

Scissors

Use office labels and stickers to decorate and seal glassine bags.

Personalize glassine bags with letter stickers.

1 Use scalloped scissors to cut scalloped edges on your washi tape, if desired. Place the washi tape on the cutting mat and use the X-ACTO knife and ruler to cut the washi tape in half lengthwise. Stick the tape on the bag at a diagonal across the corner or straight across the bag, and use scissors to trim the ends of the tape flush with the bag.

2 Add treats and seal the top of the bag with a piece of washi tape.

STITCHED FELT POUCH

Although a heartfelt gift is always number one in my book, I also enjoy giving practical gifts, such as a gift card to a friend's favorite store. Personalize a gift card by making this charming stitched felt pouch, which the recipient can then put to another use, such as to hold business cards or spare change. When cutting out this felt pouch, be sure to use fabric shears or a rotary cutter to ensure a clean cut.

continued . . .

MAKES

One felt pouch, 3 by 3¾ in/ 7.5 by 9.5 cm

MATERIALS

1 piece of wool felt, 3 by 9 in/ 7.5 by 23 cm

1 plastic button, ½ in/ 12 mm diameter

Cotton thread

Gift card, 2¼ by 3¼ in/ 5.5 by 8.25 cm

TOOLS

White marking pen

Ruler

Hand-sewing needle

Straight pins

Sewing machine

Scissors

BRIGHT IDEAS

Switch up the size and use it to wrap up any gift or as an envelope for a handmade card.

If you don't have felt, you can substitute text-weight paper or cardstock and seal with glue, tape, or a sticker.

1. Place the piece of felt vertically on a flat surface. To mark a spot for the button, use a white marking pen and a ruler to mark a point ¾ in/2 cm from one short end and 1½ in/4 cm from the sides. Using the needle and thread, sew the button on at this spot.

2. Fold the felt piece under, with 1½ in/4 cm extra material sticking out on top to create a flap. Pin into place.

3. With the sewing machine and the cotton thread, stitch ¼ in/6 mm from the edge around the entire felt piece.

4. For the button hole, use scissors to cut a ¾ in/2 cm slit in the flap, 1 in/2.5 cm from the top and 1½ in/4 cm from the sides.

5. Insert the gift card in the pouch.

6. Button the flap to close the pouch.

1

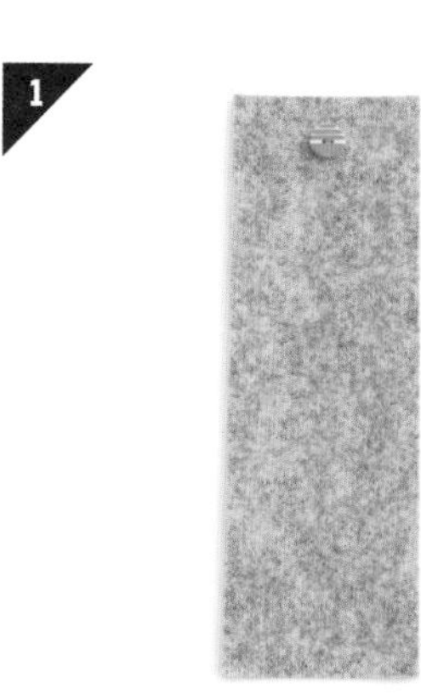

2

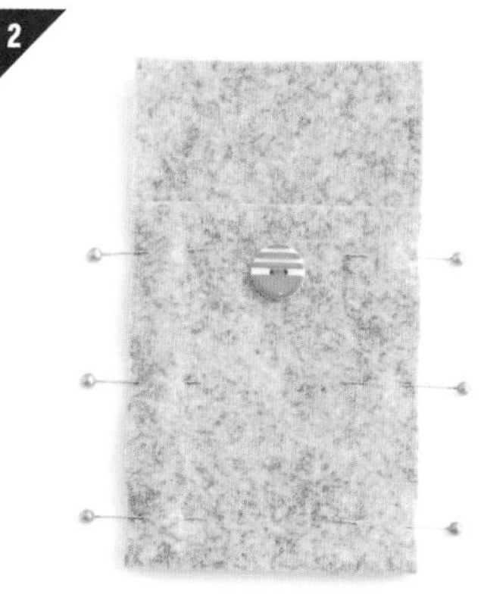

4

GIFT CARD

YARN POM-POM TOPPER

Yarn pom-poms are charming gift toppers. You can make pom-poms of different sizes using pom-pom maker kits from fabric stores, or you can make mini pom-poms by wrapping yarn around a fork and small pom-poms by wrapping yarn around your hand. Use a single pom-pom or a cluster of mini pom-poms to decorate a gift.

continued . . .

MAKES

1 pom-pom topper

MATERIALS

Ball of yarn

1 piece of yarn, 12 in/30.5 cm long

Wrapped gift

TOOLS

Scissors

BRIGHT IDEAS

Make a dozen or more yarn pom-poms and string them together to make a sweet party garland.

Try using several different colors of yarn to make a multi-color pom-pom.

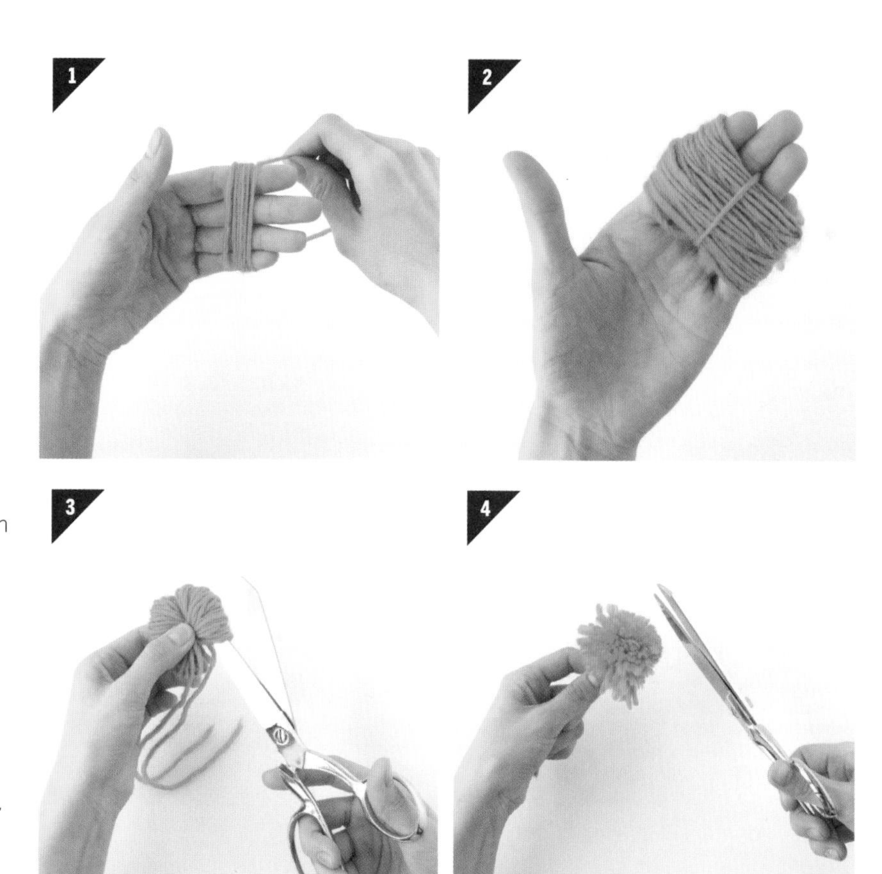

1 Place four fingers together on one hand and wrap the yarn around them fifty to sixty times (the more times you wrap the yarn, the fuller the pom-pom will be). Use the scissors to cut the end of the wrapped yarn from the ball of yarn.

2 Take the 12-in/30.5-cm piece of yarn, slip it in between your second and third fingers, wrap it very tightly around the looped yarn twice, and tie a knot.

3 Pull the yarn off your fingers and double knot the yarn. Cut through the yarn loops on each side.

4 The yarn ball will be uneven. Fluff up the yarn ball, and trim it so that it is even and spherical in shape, leaving the long yarn ends.

5 Use the yarn ends to tie the pom-pom around your gift.

FELT LETTER GIFT TOPPER

I like the idea of making a gift topper that the recipient can keep after they have opened the gift—a bonus gift! Felt letter gift toppers are easy to make and can then be hung as décor in the recipient's home. I prefer the texture of wool felt, but you can substitute with acrylic felt.

MAKES

1 felt letter

MATERIALS

1 piece of felt, 4 by 4 in/ 10 by 10 cm
Cotton thread
Twine
Wrapped gift

TOOLS

Black marking pen
Computer (optional)
Printer (optional)
1 sheet of printer paper (optional)
Scissors
Sewing machine or hand-sewing needle
Cutting mat
Leather hole punch
Mallet

BRIGHT IDEAS

Make a 3-dimensional letter by cutting out two identical letters, whipstitch around the letters leaving a 2 in/5 cm opening, stuff the 3D letter with polyfill, and slipstitch the closing. You can make the front and back letters different colors.

Instead of punching a hole, sew a loop of ribbon onto the felt letter.

1 Draw a letter onto the felt using a black marking pen. Or, if you want to use a font, select a bold font, print the letter onto printer paper, and cut the paper letter out. Then use the black marker to trace around the letter on the felt.

continued . . .

2 Use scissors to cut the letter from the felt, making sure to cut inside the black outline. With the sewing machine or a hand-sewing needle and the cotton thread, stitch ⅛ in/ 3 mm from the edge of the letter.

3 Place the felt letter on the cutting mat and use a leather hole punch and mallet to punch a hole at the top of the letter.

4 Thread the letter onto twine and attach it to your wrapped gift.

FABRIC FLAG GIFT TOPPER

Fabric flags add a festive touch to your gift, or can be placed atop cupcakes or other sweet treats. You can make these charming accessories with fabric scraps, washi tape, or paper.

MAKES

1 fabric flag

MATERIALS

1 piece of fabric, ½ by 2 ½ in/ 12 mm by 6 cm
Spray starch
Glue stick
Toothpick
Wrapped gift

TOOLS

Iron
Ironing board
Small scissors
X-ACTO knife

HOW TO

1 Spray the piece of fabric with a light mist of the spray starch. Using the iron on medium setting and no steam, iron the fabric.

BRIGHT IDEAS

Use colored masking tape, handmade washi tape (see page 111), or store-bought washi tape instead of fabric.

Paint the toothpick with acrylic paint to give it a pop of color.

continued . . .

2 Using the glue stick, apply a thin layer of glue on the wrong side of the fabric. Put the toothpick at the center of the fabric, fold the fabric over, and press down. Use the scissors to trim an inverted V-shape into the end of the fabric flag.

3 Use the X-ACTO knife to carefully make two small parallel slits on the gift-wrapped box. Slide the toothpick through the two slits.

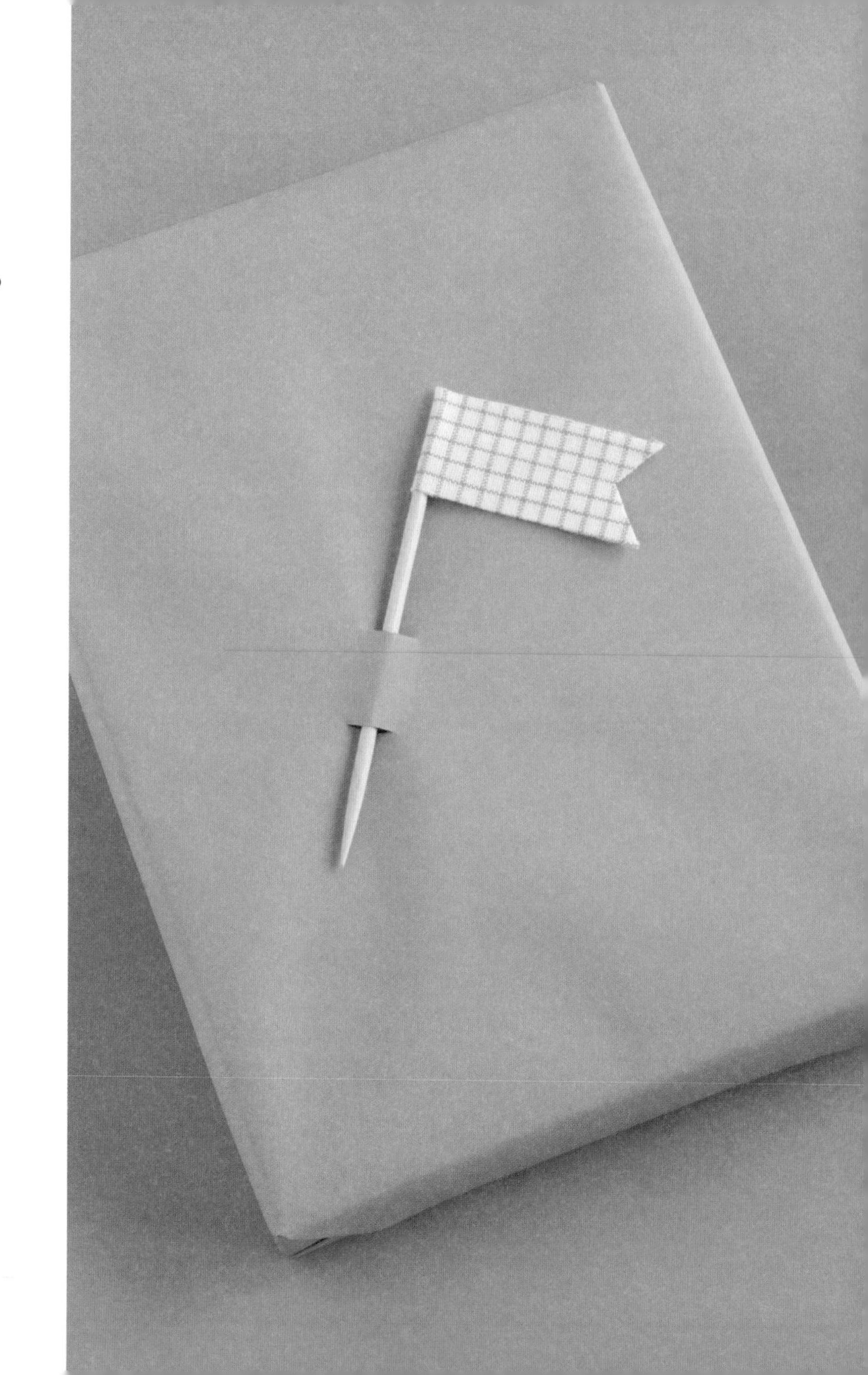

WOOD VENEER BOW GIFT TOPPER

Wood veneer is a fun material to work with and provides texture and visual interest to any gift. You can find wood veneer at woodworking suppliers or at some paper stores. To bend the wood veneer without cracking it, you can either use steam or place it in water. The wood veneer bow takes a bit of time to make because of the drying time, but it is worth the effort. You can make several wood bows at one time and store them for future gift-wrapping projects.

MAKES

1 bow

MATERIALS

Two pieces of wood veneer; one 1 by 6 in/2.5 by 15 cm, one ½ by 2 in/12 mm by 5 cm

Wood glue

Wrapped gift

TOOLS

Bowl of hot water

Towel

Clothespins

1 Place the two wood veneer pieces in the bowl of hot water. Let them soak for 2 to 3 minutes. Use the towel to pat them dry. Bend the ends of the larger wood piece under to form the main bow shape, so that the ends meet in the middle of the bow.

continued . . .

Use wool felt instead of wood veneer to make a bow.

After you make your bow, use a paint pen to write a message on the wood veneer.

2. Bend the smaller piece of wood veneer so that it wraps around the larger wood piece and overlaps the middle of the bow.

3. Use the clothespins to hold the wood veneer pieces in this shape. Let the wooden bow dry for 24 hours.

4. Remove the clothespins and place a dab of wood glue on the inside of the bow to reinforce it. Put a dab of wood glue on the smaller piece of wood veneer to secure where it overlaps the bow. Put clothespins on the wooden bow again to hold it in place. Let the glue dry for 24 hours.

5. Remove the clothespins and attach the wood veneer bow topper to your wrapped gift using wood glue. Let dry for 1 hour.

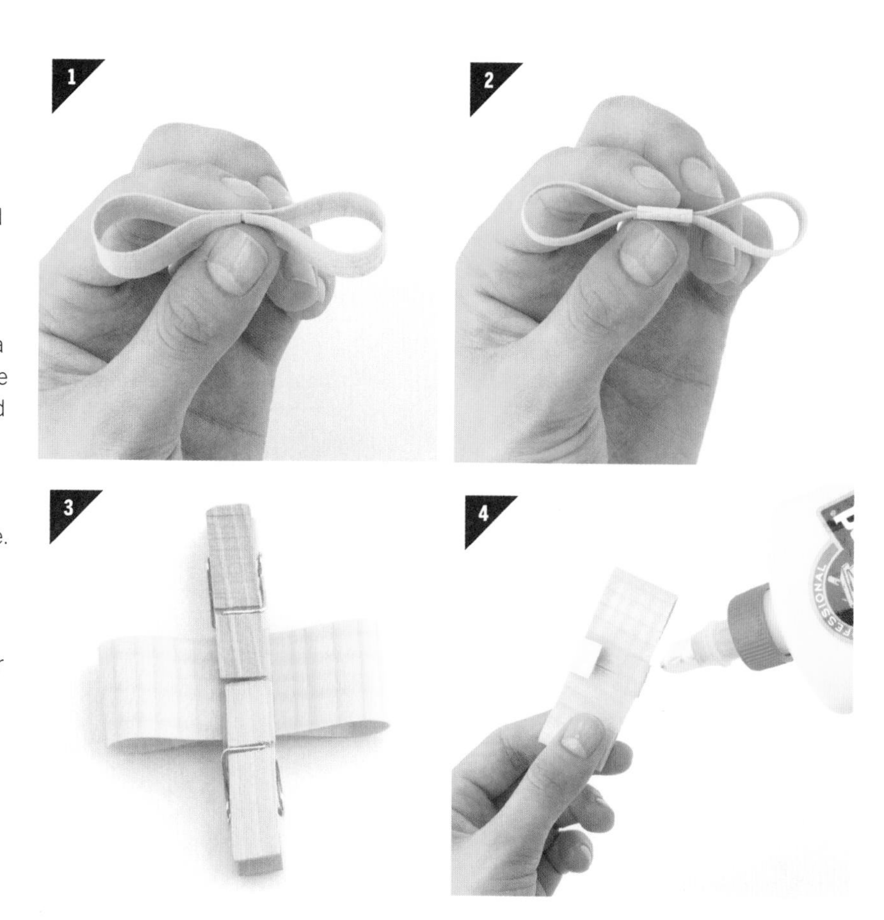

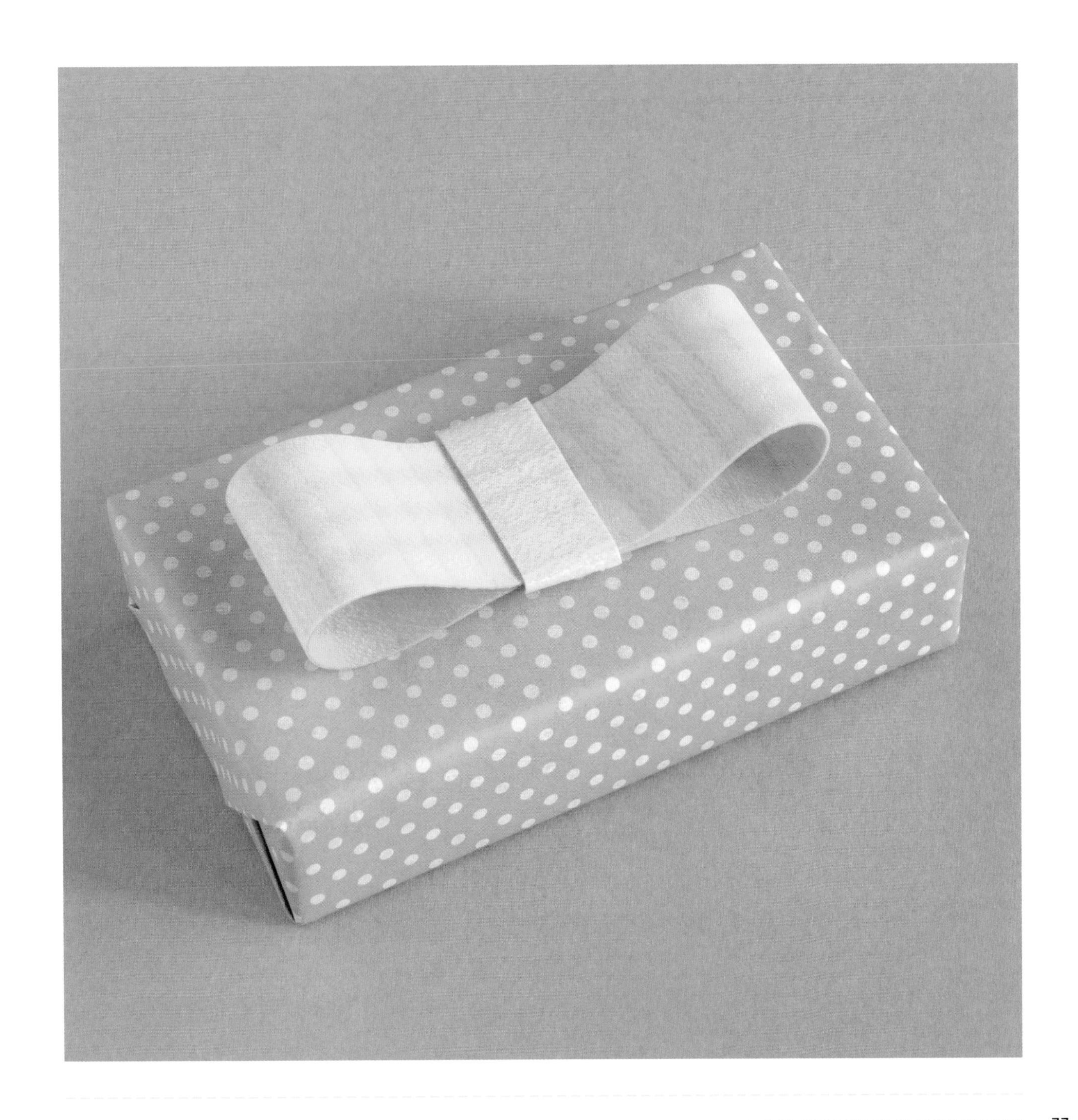

FABRIC BUTTON GIFT TOPPER

MAKES

1 button topper

MATERIALS

1 piece of cotton fabric, 1¾ by 1¾ in/4.5 by 4.5 cm

1 hair elastic, 1¼ in/3 cm in diameter

Wrapped or boxed gift

Washi tape (optional)

TOOLS

One ¾-in/2-cm cover button set with pusher tool and circle template

Pencil

Scissors

Pointed tweezers

BRIGHT IDEAS

Instead of adding an elastic ring to the back of the button, string some twine through the loop and wrap the twine around your gift.

Use a piece of twine to string multiple fabric buttons to create a button garland for your gift.

These charming fabric-covered button gift toppers can be reused as fancy rubber band substitutes or as hair ties. They are easy to make, so keep a bunch handy in your craft stash to add a personal touch to any gift. Fabric with a small print works best for this project. Fabric button kits include buttons, mold, and pusher. The kits are inexpensive and can be found in the notions section of fabric shops.

1. Use the circle template and pencil to trace a circle onto the fabric. Cut out the circle with the scissors.

2. Center the fabric circle right-side down on the plastic button form and place the button shell face down on the fabric. Push the shell down into the button form and tuck the fabric edges neatly into the well of the button shell.

3 Place the button back on top of the button shell. Use the pusher tool to firmly press the back into the shell.

4 Remove the plastic pusher tool from the button and the button from the button form. Use the pointed tweezers to pull a small portion of the hair elastic through the button shank (the metal loop) so the loop divides the elastic into two parts. Insert the longer part of the elastic through the small elastic loop and pull it until it is secure.

5 Place the fabric button gift topper around a gift box. If you'd like, tape a strip of washi tape on the package under the button for an even more decorative punch.

NOTE: If you are using a large box, use a piece of elastic string to secure the fabric button around the gift.

TISSUE PAPER FLOWER GIFT TOPPER

Tissue paper flowers make charming additions to any gift. You can recycle tissue paper from previous gifts you've received or keep a stock of multicolored tissue paper packs. These flowers can be monochromatic or you can use tissue paper of different colors in one flower.

continued . . .

MAKES

1 flower topper

MATERIALS

- Eight 5-in/12-cm squares of tissue paper (same color or a variety of colors)
- 1 piece of floral wire, 6 in/15 cm long
- Craft glue
- Wrapped gift

TOOLS

Scissors

BRIGHT IDEAS

Create a different flower by cutting the edges in a point.

Twist the wire around a piece of ribbon and secure it to the wrapped gift.

1. Place the eight squares of tissue paper in one stack. Fold the stacked squares of tissue paper accordion-style using approximately 1-in/ 2.5-cm folds.

2. Fold the floral wire in half and wrap it tightly around the center of the folded paper. Twist the wire several times to secure it in place.

3. Using the scissors, cut rounded ends on the folded paper.

4. Gently unfold one side of the flower by separating the tissue paper layers from each other. Repeat on the other side. Fluff the petals to increase the fullness of the flower.

5. Clip the wire ends with scissors, and glue the tissue paper flower on your wrapped gift.

PAPER ROSETTE GIFT TOPPER

Paper rosettes provide a decorative pop of color and shape to any gift. You can make one large rosette to fit the entire top of the gift box or you can cover the gift box with a cluster of small paper rosettes. The recipient can hang the lovely rosette as a decoration or use it for a future gift-wrapping project.

1 Lay the text-weight paper vertically on the cutting mat. Using the ruler as your guide, score the paper vertically with the bone folder at ½-in/12-mm intervals.

continued . . .

MAKES

1 rosette topper

MATERIALS

1 sheet of text-weight paper, 8½ by 11 in/21.5 by 28 cm

Double-sided tape

1 piece of cardstock, 1 by 1 in/2.5 by 2.5 cm, in a coordinating color

Wrapped gift

TOOLS

Cutting mat

Ruler

Bone folder

Scissors

X-ACTO knife

Paper punch, ½-in/12-mm circle

BRIGHT IDEAS

Make several different paper rosettes in different colors and sizes and layer them.

To make a larger rosette, fold two pieces of paper (do not cut the paper in half), and attach according to the instructions.

2 Turn the paper horizontally, and fold along each scored line accordion style to make pleats. If you would like the edges of each pleat to be rounded or pointed, cut each end into a U- or V-shape.

3 Open up the paper and cut down the center with the X-ACTO knife so that you have two 8½-by-5½-in/21.5-by-14-cm accordion-folded pieces of paper.

4 On one piece of paper, squeeze the pleats together and fold in half, end to end. Place a strip of the double-sided tape along one side of the interior fold, to join the two halves.

5 Press together and you will have one half of the rosette. Repeat steps 3 and 4 with the other piece of paper to make the other half of the rosette.

6 Fan out the pleats on both pieces of paper. Place double-sided tape along the edge of one half of the rosette and press together with the other half to form one rosette.

7 Using the circle paper punch, punch a ½-in/12-mm circle from the piece of cardstock. Place a small piece of the double-sided tape on the back of the cardstock circle, and stick it in the center of the rosette. Use a piece or two of double-sided tape to stick the paper rosette to the top of the wrapped gift.

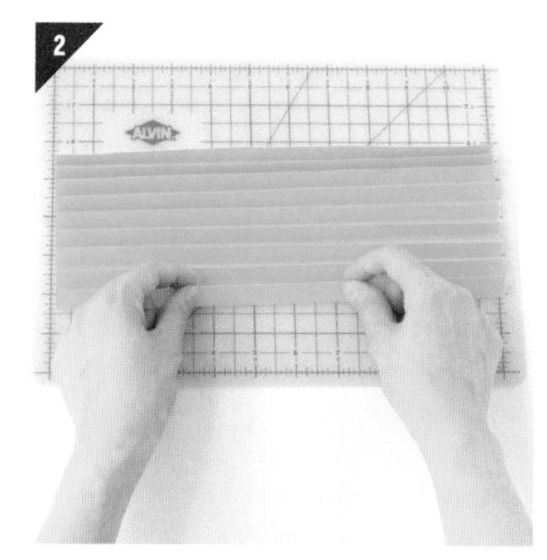

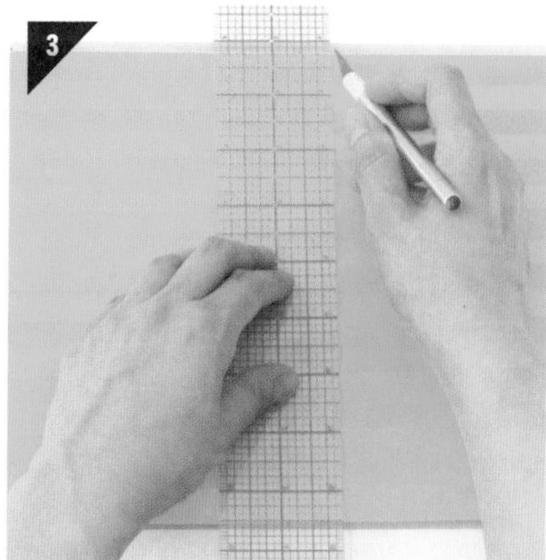

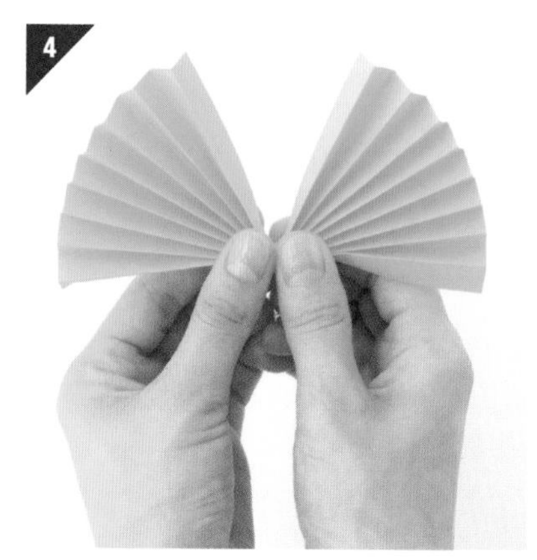

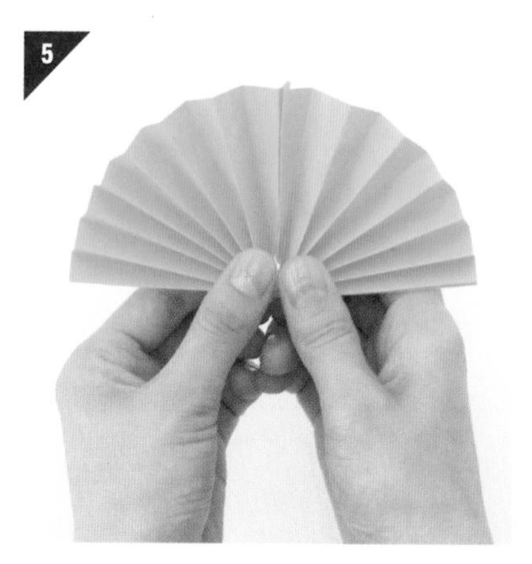

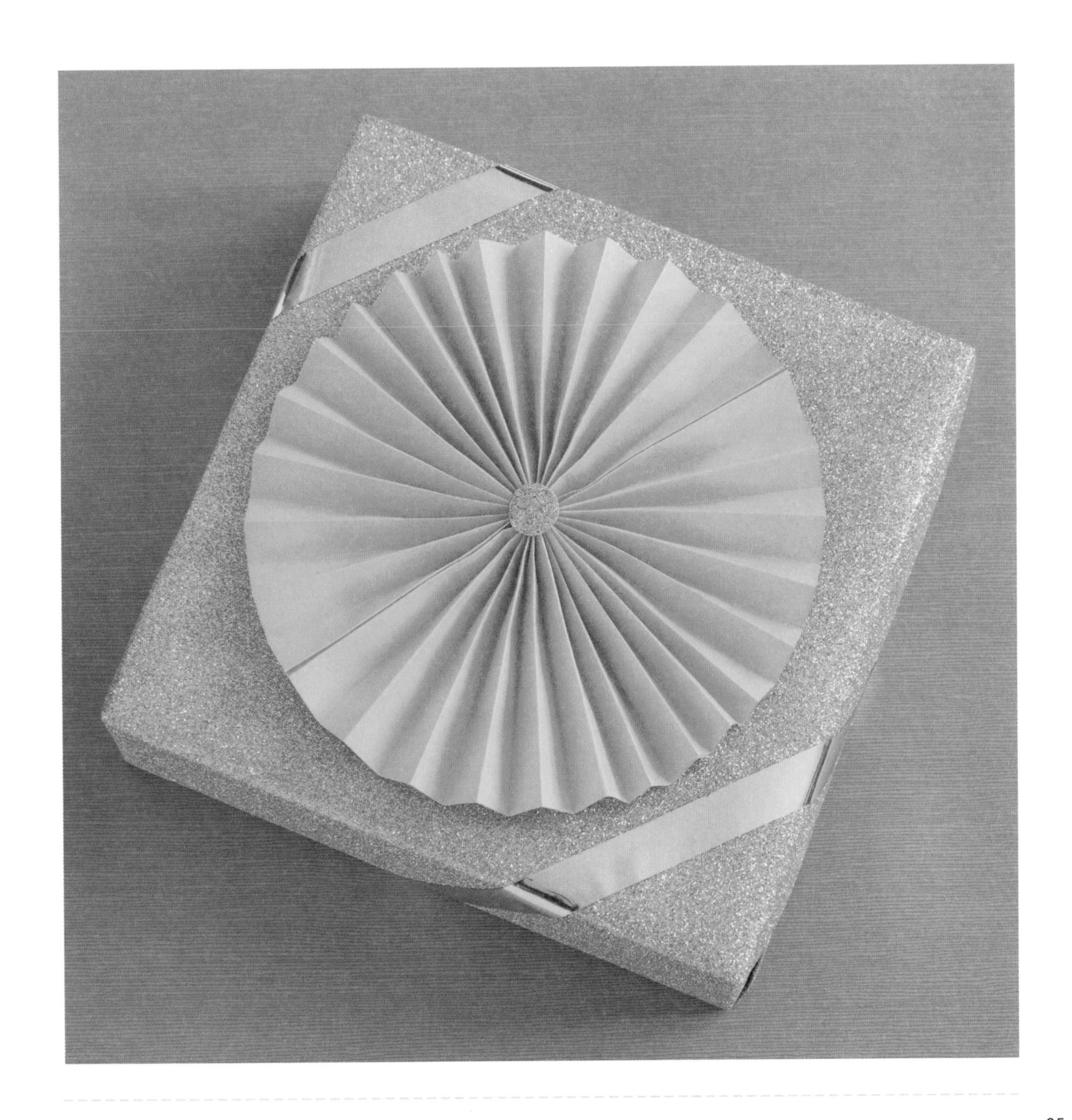

STITCHED PAPER GARLAND

MAKES

One 25-in/65-cm garland

MATERIALS

Gold and silver cardstock

Cotton thread

Tape

TOOLS

½-in-/12-mm-diameter circle paper punch

Sewing machine or hand-sewing needle

Scissors

Garlands are a fun way to dress up a simple package. You can use colored cardstock, patterned scrapbook paper, vellum, or wood veneer. The garland can be made using any shape of paper punch. It can be reused on a package or put up as decoration.

BRIGHT IDEAS

Use a paper punch to cut out old photographs and stitch them together to make a garland.

Cut up small pieces of fabric and stitch them together to make a garland.

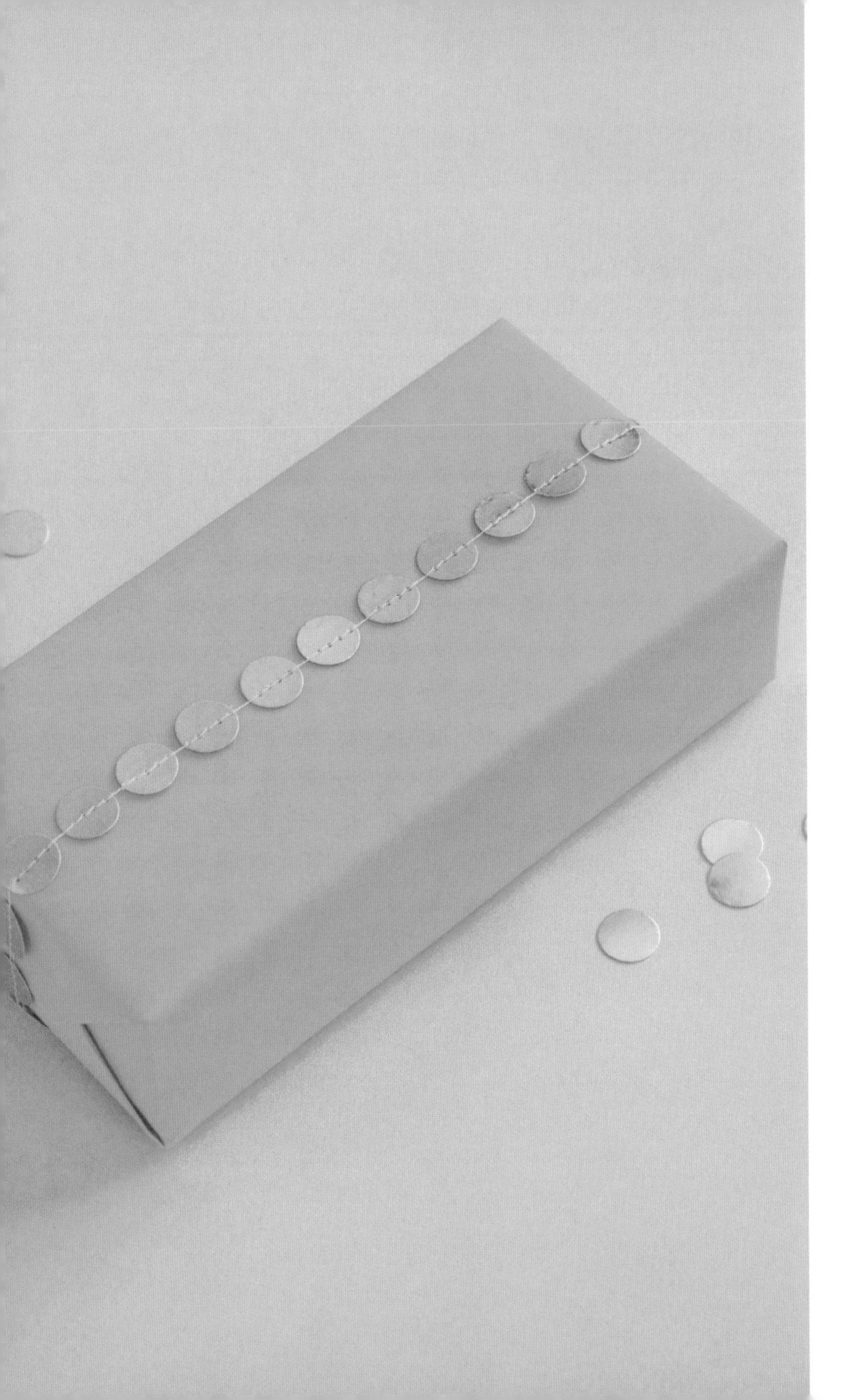

1 Using the circle paper punch, make twenty-eight gold and silver paper circles. Lay out the paper circles in a linear color array of your choice to make a garland.

2 Thread the sewing machine with the cotton thread. Pull out a 4-in-/10-cm-long thread tail from the top and bobbin threads before you begin to sew the garland. Feed the paper circles through the sewing machine, leaving a ⅛-in/3-mm gap between each circle. After sewing through the last circle, leave another 4-in-/10-cm-long thread tail.

3 Wrap the paper circle garland around the gift and secure it with tape. Clip off the remaining thread tails.

FELTED BALL GARLAND

MAKES

1 garland of 5 or 6 felted wool balls

MATERIALS

Wrapped gift

A few handfuls of wool roving

Cotton embroidery thread

TOOLS

1 tbsp dishwashing soap

Small bowl containing 1 cup/240 ml hot water

Spoon

Small bowl of cold water

Embroidery needle

Crafting felted balls requires few materials and is a great addition to a gift when you have a little time on your hands. All you need is wool roving, dishwashing soap, and water and you can make felted balls of all sizes. Wool roving can be found at knitting or craft stores. If you don't have time to make felted balls, you can use store-bought mini pom-poms.

1 Add the 1 tbsp dishwashing soap to the small bowl with 1 cup/240 ml of hot water. Stir the dishwashing soap and hot water with a spoon until it is mixed well. Tear off a small handful of wool roving (approximately one-eighth of the total roving); the ends will be frayed when you make the tear. Gently dip the wool into the soapy water. You want the wool to be slightly damp, but not fully wet.

continued . . .

Instead of stringing the felted balls, you can glue individual felted balls in a random pattern on the wrapped box.

Mix two or more colors of wool roving to create multicolored felted balls.

2 Gently rub the wool between your hands, forming a ball shape. Once a ball shape is formed, dip the wool into the small bowl with cold water.

3 Continue to rub the wool while adding more pressure.

4 Alternate between rubbing the wool and dipping the wool into the hot soapy water and the cold water until the ball feels firm and dense. Repeat this process to make the desired number of felted balls. Let the felted wool balls dry for 24 hours.

5 Thread the needle with the embroidery thread and use it to string the felted balls, leaving enough of the thread at each end to wrap around the gift and secure with a knot. Trim the thread ends.

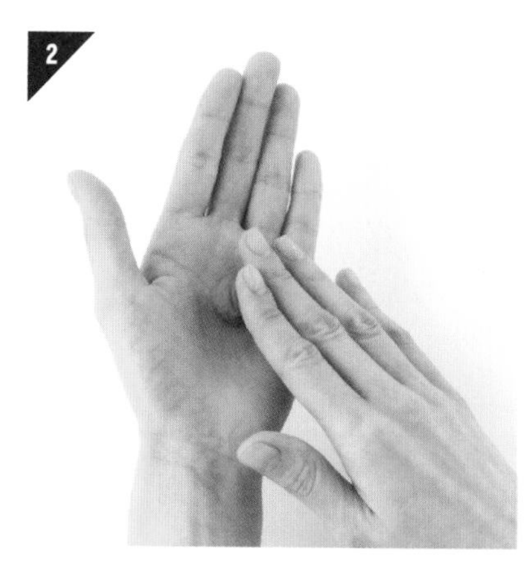

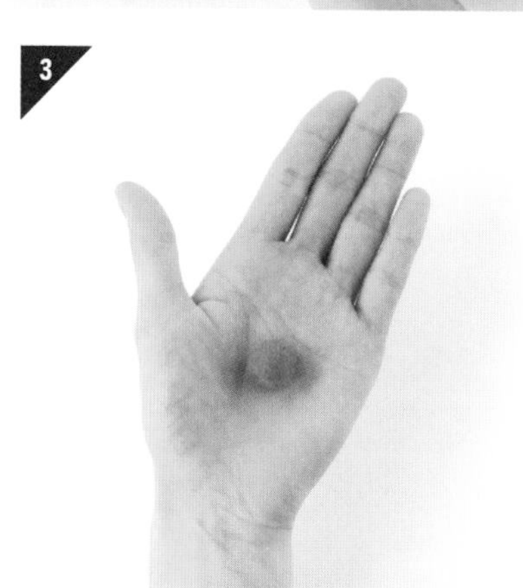

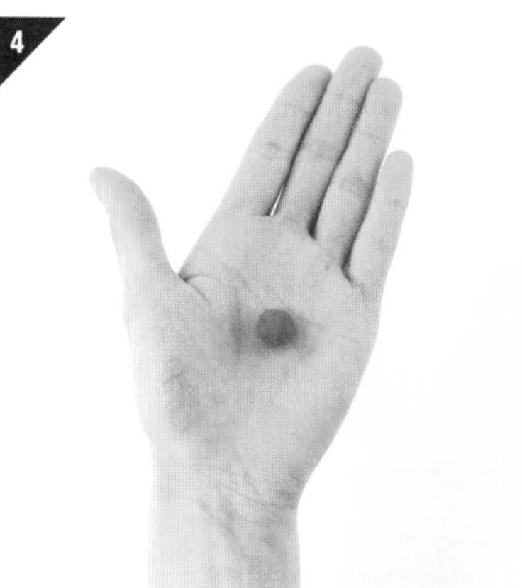

BROWNIE MIX GLASS JAR KIT

I enjoy giving edible gifts to friends because I know they won't sit around and collect dust. Not only can the recipient enjoy a treat, but they can reuse the glass container. Brownie mixes are a sweet treat that everyone will enjoy, especially when you've done all the hard work for them (you'll be providing the ingredients, but don't mix them up!). Including a stamped wood spoon and a mini recipe card adds a warm touch.

BRIGHT IDEAS

Make s'mores kits using mini marshmallows, graham crackers, and chocolate.

Use glass paint pens to decorate the glass container.

continued . . .

MAKES

1 jar

MATERIALS

One 36-oz/1-L clean, dry glass jar with lid

3¾-in-/9.5-cm-long wooden spoon

1 manila tag, 2½ by 4 in/6 by 10 cm

1 piece of twine, 12 in/30.5 cm long

INGREDIENTS FOR BROWNIE MIX

1 cup/200 g granulated sugar

1 cup/200 g firmly packed brown sugar

½ cup/60 g all-purpose flour

½ tsp salt

1¼ cups/120 g cocoa powder

6 oz/170 g chocolate chips (optional)

½ cup/55 g chopped nuts (optional)

TOOLS

Rolling alphabet stamp

Black stamp pad

Drill with ⅛-in/3-mm wood drill bit

Scissors

1 Pour half of the granulated sugar into the glass jar, followed by all of the brown sugar. Next, pour in the remaining granulated sugar, followed by half of the flour and salt. Finally, pour in all of the cocoa powder, and top it off with the remaining flour and salt. If you have extra space at the top of your container, add some of the chocolate chips or nuts, if desired. Put the lid on the glass jar.

2 Using the rolling alphabet stamp (a stamp with individual dials with the entire alphabet) and the black stamp pad, stamp "BROWNIES" on the spoon handle. Drill a hole in the end of the handle.

3 On the manila tag, write or print out the recipe using the following text to include with the jar. (If you don't have a manila tag you can make your own gift tag using the instructions on page 98.)

4 String the twine through the holes in the spoon and the recipe card, and wrap the twine around the neck of the glass jar. Secure the twine with a knot and trim the ends with the scissors.

RECIPE CARD

Makes 16 brownies

INGREDIENTS

4 eggs

1 glass jar brownie mix

8 oz/225 g butter, melted

2 tsp vanilla extract

DIRECTIONS

- Preheat the oven to 300°F/150°C. Butter and flour an 8-in/20-cm square pan.
- Using an electric mixer, beat the eggs at medium speed until fluffy. Add the brownie mix and beat to combine. Add the melted butter and the vanilla and beat to combine.
- Pour the batter into the prepared pan and bake for 45 minutes. Place the pan on a rack to cool for 1 hour.
- Cut into 2-by-2-in/5-by-5-cm squares. Store in an airtight container at room temperature for up to 4 days.

BROWNIES
Ingredients
4 eggs
Directions

BAKED GOODS PACKAGING

Parchment paper is not only handy to use when baking, but it also makes a lovely natural base wrap for baked goods. Add some visual interest by adding a simple fabric wrap using a scrap from your stash. Fat quarters, sold in fabric stores, are the perfect size to make fabric wraps.

MATERIALS

(TO WRAP ONE 9-BY-5-BY-3-IN/23-BY-12-BY-7.5-CM LOAF)

- 1 piece of parchment paper, 8 by 17 in/20 by 43 cm (or 13 by 18 in/33 by 46 cm if wrapping entire baked good)
- Masking tape
- 1 piece of fabric, 7 by 17 in/17 by 43 cm
- 1 piece of linen ribbon, 20 in/50 cm long

TOOLS

- Scissors

BRIGHT IDEAS

Instead of using a fabric wrap, use strips of washi tape to seal the parchment paper.

Use colored glassine paper or freezer paper to package baked goods.

1 Wrap the parchment paper around the baked item and seal with masking tape on the underside. For immediate delivery, you can leave the ends of the baked item exposed. If you will not be delivering it right away, wrap the entire baked good in parchment paper.

2 Wrap the piece of fabric around the center of the baked good and seal with masking tape on the underside.

3 Wrap the ribbon around the fabric and tie in a knot. Use scissors to trim the ends of the ribbon so they are an even length.

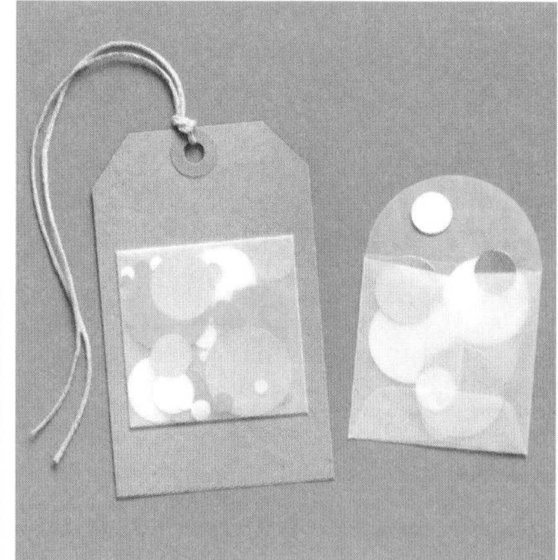

PROJECTS **PART 3**

GIFT TAGS AND RIBBON

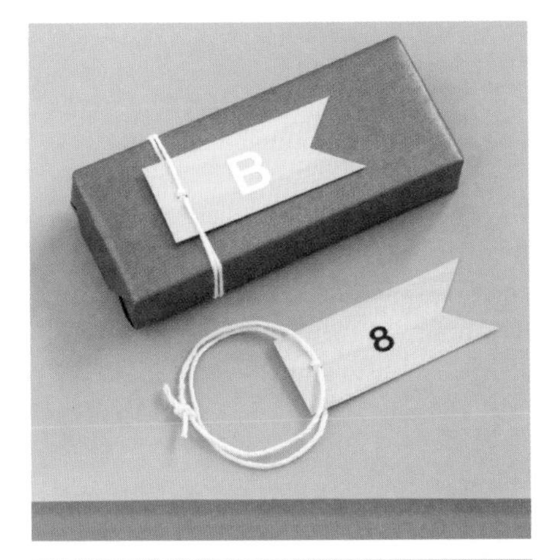

I love visiting bakeries and peeking through to the back room to see the patissier putting the final touches on cakes. Adding a handmade gift tag or ribbon is like icing on the cake; it's the final creative element that adds an extra detail to your gift. Personalize your packages with the ideas in this section.

33 DIY Gift Tags
34 Confetti Gift Tag
35 Paint Chip Gift Tag
36 Wood Gift Tag
37 Personalized Leather Gift Tag
38 Clay Gift Tag
39 Handmade Patterned Washi Tape
40 Hand-Painted Patterned Twill Tape
41 Hand-Dyed Twill Tape
42 Painted Wooden Bead Ribbon Ends
43 Linen Ribbon
44 Custom Paper Envelope
45 Nail Polished Brads

DIY GIFT TAGS

MAKES

6 gift tags

MATERIALS

6 pieces of cardstock, 2 by 3½ in/5 by 9 cm, in selected colors

Paper hole reinforcement stickers

Washi tape in various colors and patterns

Gold foil tape

Letter stickers

TOOLS

Cutting mat

Pencil

X-ACTO knife

Ruler

Hole punch

Rubber stamps

Rubber stamp pad

You can make your own charming gift tags using cardstock, a paper hole punch, and paper hole reinforcement stickers. Embellish the tags with washi tape and gold foil tape, which adds a bright metallic pop. Letter stickers, other kinds of stickers, and rubber stamps are also simple but fun ways to enhance sweet gift tags.

BRIGHT IDEAS

Use watercolor to paint an abstract pattern onto a manila gift tag.

Stamp a custom design on a manila tag to make a personalized gift tag.

1 Lay a piece of the cardstock vertically on the cutting mat. Using the pencil, make a mark on the edge of the cardstock ½ in/12 mm in from one top corner and another mark ½ in/12 mm down from the same corner. Repeat on the opposite top corner. Using the X-ACTO knife and the ruler, connect these two marks and cut off the corners.

2 In the middle of the top of the gift tag, make a mark with the pencil ½ in/12 mm from the edge. Using the hole punch, make a hole at this mark, and add reinforcement stickers over the hole on both sides of the gift tag.

3 Repeat steps 1 and 2 to make the five remaining gift tags.

4 Decorate the gift tags using washi tape, gold foil tape, letter stickers, and rubber stamps.

CONFETTI GIFT TAG

MAKES

1 gift tag

MATERIALS

Scraps of colored paper in a variety of sizes and colors

1 glassine envelope, 2 by 2 in/ 5 by 5 cm

Double-sided tape

1 gift tag, 2½ by 4 in/ 6 by 10 cm

Twine

TOOLS

Paper hole punch

Paper circle hole punches in a variety of sizes

Confetti epitomizes fun festive occasions. Homemade confetti is easy to make and is a clever way to use up scraps of colored paper. Create these simple confetti gift tags to give to friends to brighten up their day. If you don't have time to make your own confetti, pick up a pack of confetti at a party supply store.

BRIGHT IDEAS

Instead of confetti, place colored candy sprinkles inside the glassine bag.

Create a birthday party invitation by printing the party details on a card and adhering the glassine bag of confetti to the invitation.

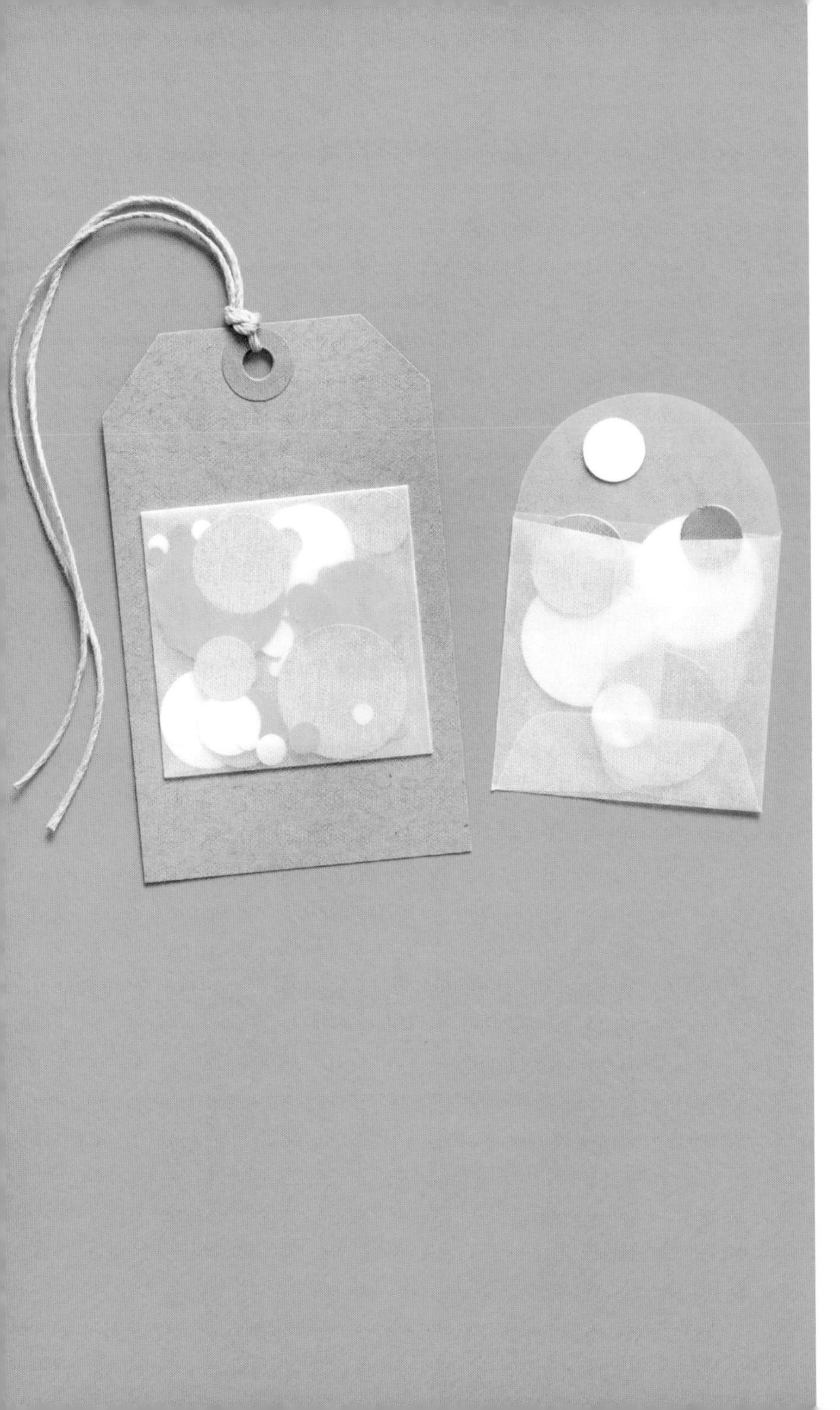

HOW TO

1. If creating your own tag, follow steps 1 and 2 on page 99, using a 2½ by 4 in/6 by 10 cm piece of cardstock.
2. Using the paper hole punch and circle hole paper punches, punch at least twenty circles from the colored paper scraps to make confetti.
3. Put the paper confetti in the envelope and seal the envelope closed with a piece of double-sided tape. Place four strips of tape on the back of the sealed envelope, one on each edge.
4. Place the glassine envelope at the center of the gift tag, and press down to adhere it to the gift tag.
5. Thread the twine through the hole and secure in a knot.

MAKES

1 gift tag

MATERIALS

Paint chip card
Double-sided tape

TOOLS

Cutting mat
X-ACTO knife
Ruler
Bone folder
Paper hole punch, ⅛-in/3-mm hole
Twine

PAINT CHIP GIFT TAG

I love perusing the aisles of home improvement stores, particularly the paint section, where I find inspiration looking at the colorful palettes of the paint chip samples. I often take home a handful of my favorite color combinations. In a few simple steps, paint chip cards can be transformed into lovely gift tags.

BRIGHT IDEAS

Use index or library cards and make them into gift tags.

Use old photographs and make them into gift tags.

1. Place the paint chip card on the cutting mat. Using the X-ACTO knife and the ruler, cut the paint chip card into a 2-by-6-in/5-by-15-cm rectangle.

2. Put double-sided tape on the wrong side of the paint chip card and use the bone folder to fold the paint chip card in half, wrong-sides together.

3. Punch a hole at one end of the rectangle, 3/8 in/1 cm from the edge.

4. Thread the twine through the hole and secure in a knot.

WOOD GIFT TAG

MAKES

1 gift tag

MATERIALS

1 piece of wood veneer, 1 by 2½ in/2.5 by 6 cm

1 letter sticker, ½ in/12 mm

Twine

Wrapped gift

TOOLS

Cutting mat

X-ACTO knife

Ruler

Paper hole punch, ⅛-in/3-mm hole

Scissors

BRIGHT IDEAS

Use a paint pen to write on the wood gift tag.

Use a rub-on transfer to apply a number or letter to the wood gift tag. Apply a coat of clear varnish on top.

Here is a tag the recipient will want to reuse or keep forever. Wood gift tags make the keeper list because they are pretty and unique. If you want a pop of color, you can paint the wood prior to decorating it.

1 Put the piece of wood veneer on the cutting mat and use the X-ACTO knife and the ruler to cut out a V-shaped notch on one end.

2 With the paper hole punch, punch a hole at the opposite end of the rectangle, 3⁄8 in/1 cm from the edge.

3 Place the letter sticker in the center of the wood veneer.

4 Thread the twine through the hole, secure it in a knot, and trim the ends of the twine with the scissors.

5 Attach the wood veneer gift tag to the wrapped gift.

MAKES

1 gift tag

MATERIALS

1 piece of leather, 1½ by 2½ in/4 by 6 cm

Twine

Wrapped gift

TOOLS

Sponge

Water

Cutting mat

Leather hole punch

Mallet

Alphabet stamp set for leather

Scissors

You can paint the leather with leather paint.

Use the alphabet stamp set on clay.

PERSONALIZED LEATHER GIFT TAG

Alphabet stamp sets for leather are one of my favorite things to work with. You can make gift tags using scraps of leather, which you can find at fabric or leather shops. Then decorate the tags with the person's name, initials, or a special date. If you don't have any leather handy, use the stamp set on chipboard or cardstock for a blind embossed look.

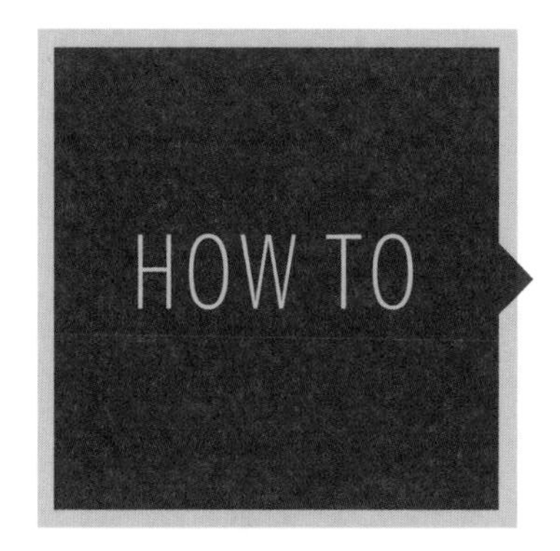

1 Dampen the sponge with water and wring out any excess water. Wipe the leather piece on both sides with the damp sponge. (The water makes the leather softer and easier to work with.)

2 Put the leather piece on the cutting mat. Place the leather hole punch centered along one short end, ¼ in/ 6 mm from the edge. Hit the leather punch with the mallet to create a hole.

3 Put the leather piece on a firm work surface. Using the alphabet stamp set and mallet, write out the recipient's name or initials in the center of the leather. (Test out the alphabet stamp on a scrap piece of leather in advance, so you know how hard you need to hit the mallet to get the desired impression.) Start from the middle letter and work your way out.

4 Tie a piece of twine around the package and secure with a knot. Thread one end of twine through the leather tag hole, secure with a knot, and trim.

2015
2015.01.25
SJS

CLAY GIFT TAG

These clay gift tags give your gift a personalized feel and make cherished keepsakes. Air-dry and polymer clays are both malleable and ideal to work with for craft projects. Air-dry clay hardens when exposed to air over a period of time and polymer clay hardens when baked in an oven. You can make these tags in a variety of sizes and they can be reused as a holiday ornament or put on a long string to make a necklace. Rubber or metal alphabet stamps work well for imprinting initials, a date, or a special message onto the clay gift tag.

continued . . .

MAKES

1 gift tag

MATERIALS

Small ball of air-dry or polymer clay, about 1 in/ 2.5 cm in diameter

TOOLS

1 piece of parchment paper, 8 by 12 in/20 by 30.5 cm

Rolling pin

1 circle cookie cutter, 1 in/ 2.5 cm in diameter

Alphabet rolling stamp

Wooden skewer

If you want to give your clay gift tags some color, let the clay dry for 24 hours and then use acrylic paint to paint the clay gift tag. Let the paint dry for 24 hours.

Try a variety of cookie cutter shapes and sizes.

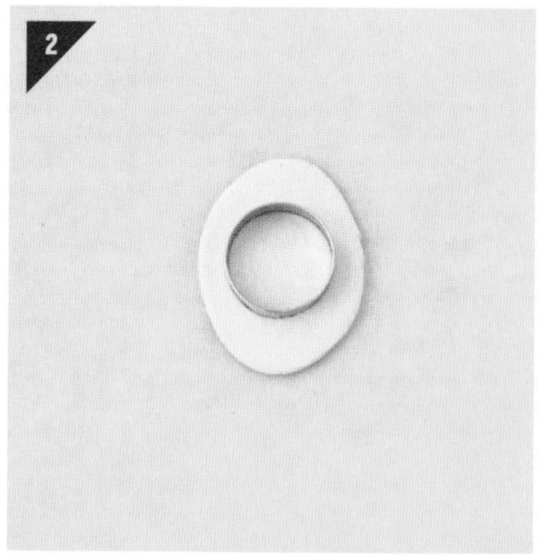

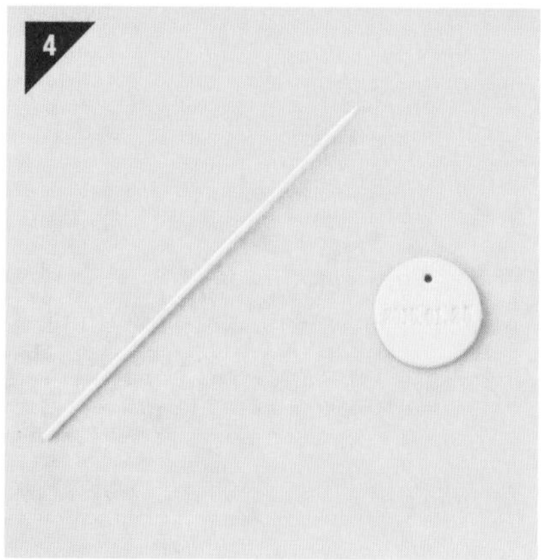

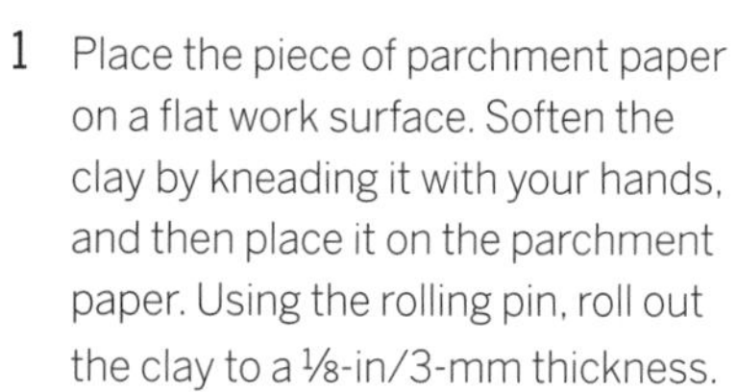

1. Place the piece of parchment paper on a flat work surface. Soften the clay by kneading it with your hands, and then place it on the parchment paper. Using the rolling pin, roll out the clay to a ⅛-in/3-mm thickness.

2. Cut out a 1-in/2.5-cm circle from the clay using the cookie cutter.

3. Gently press the alphabet rolling stamp onto the clay circle. (I recommend practicing stamping on a sample of clay so you know how hard to press on the stamp to obtain a clear impression in the clay.)

4. Press the end of the wooden skewer into the clay to create a hole to hang the clay gift tag. Let the clay dry for 24 hours. (If using polymer clay, bake the circle in the oven according to the manufacturer's instructions.)

HANDMADE PATTERNED WASHI TAPE

Washi tape is one of those must-have staples for every crafter's toolbox. Since washi tape can be repositioned easily, it's easy to work with and is versatile in application. To transform a box into something elegant, make your own washi tape using medical paper tape (found at your local pharmacy) and gel pens.

MATERIALS

Medical paper tape

Wrapped gift

TOOLS

Scissors

Cutting mat

Gel pens in a variety of colors

HOW TO

1 Use scissors to cut a strip of the medical paper tape long enough to wrap around your wrapped gift, plus 1 in/2.5 cm. Place the paper tape on a cutting mat and use the gel pens to create a pattern, to make your own "washi" tape.

2 Put your handmade washi tape around the gift, overlapping the ends on the underside of the gift.

Use scalloped scissors or pinking shears to create scalloped or zigzag washi tape.

Cut the washi tape into pieces to create decorative geometric shapes.

MATERIALS

- 1 piece of 100% cotton white twill tape, 1 in/2.5 cm wide and 18 in/46 cm long
- Fabric paint (in as many colors as you like)
- Wrapped gift

TOOLS

- Piece of scrap paper to place under the twill tape
- Masking tape, ½ in/12 mm wide
- Shallow plastic bowl
- Foam paintbrush
- Iron
- Ironing board

Hand-sew or machine stitch on the cotton twill tape to create stripes or a pattern.

Use a rubber stamp and a fabric ink stamp pad to stamp a pattern on the twill tape.

HAND-PAINTED PATTERNED TWILL TAPE

Twill tape provides the perfect base for creating a patterned ribbon. You can use masking tape and fabric paint or rubber stamps and a fabric ink stamp pad to create patterns on plain twill tape. Smooth cotton twill tape works best for this project.

1 Place the twill tape on top of the scrap paper and use a small piece of masking tape to secure each end of the twill tape to the scrap paper.

2 Tear off small pieces of masking tape, and place them diagonally across the twill tape, along the entire length of the twill piece, and press down to ensure the tape is secure. Make sure the tape pieces go over the edges of the twill tape, and vary the widths between masking tape pieces.

3 Put a small amount of fabric paint in the plastic bowl. Dip the foam paintbrush into the paint, and remove excess paint on the edge of the bowl. Paint on the exposed sections of the twill tape, making sure to apply an even coat of paint. Let the paint dry for 24 hours. Depending on the color of the paint, you may choose to paint a second coat.

4 Carefully remove all the masking tape from the twill tape.

5 Turn on the iron to medium-high heat with no steam. Place the twill tape painted-side down on the ironing board and heat set the fabric paint by ironing the back of the twill tape for 2 to 3 minutes. Make sure to apply even pressure over the entire length of the twill tape.

6 Wrap the twill tape around the gift, painted-side out, using the instructions on page 16.

HAND-DYED TWILL TAPE

Among the wonderful selection of ribbons to be found, many are quite pricey. Twill tape is a fantastic alternative to premium ribbon. I love using twill tape to wrap gifts because it is inexpensive and comes in a wide array of widths and textures. Although I love white or natural color twill tape, you can brighten it up by dyeing it a solid color or creating an ombré or tie-dye effect.

continued . . .

MATERIALS

2 tbsp/30 ml powdered fabric dye

1 cup/240 ml hot water

1 yd/1 m of 1-in-/2.5-cm-wide 100% cotton white twill tape

Wrapped gift

TOOLS

Plastic gloves

Shallow heat-resistant bowl

Wooden popsicle stick

Small metal tongs

12 to 16 rubber bands of various sizes (optional)

If you don't have twill tape handy, try using white cotton shoelaces or strips of light-colored cotton fabric.

After dyeing the twill tape one color, let it dry and then place the twill tape in another colored dye solution for a multicolor effect.

1 Put on the plastic gloves and prepare the dye in the bowl following the manufacturer's instructions.

2 To dye the entire piece of twill tape, using the metal tongs, place the twill tape into the dye solution and stir it around while it soaks up the dye. Once the twill tape becomes the desired shade, remove it from the dye solution and rinse it under cold water for 1 minute. Let the twill tape dry for 24 hours.

3 To create an ombré effect in the piece of twill tape, roll up the tape and dip two-thirds of the tape roll into the dye. Rinse under cold water. To create graded color portions, dip the smaller portion of the tape for a longer period of time. Once the desired ombré effect is achieved, remove the twill tape from the dye solution and rinse under cold water for 1 minute. Let the tape dry for 24 hours.

4 To create a tie-dye effect, if desired, wrap a dozen or more small rubber bands tightly at different intervals around the twill tape before placing it into the dye solution. The thicker the rubber bands, the wider the tie-dye stripes will be. Once the twill tape becomes the desired shade, remove it from the dye solution and rinse under cold water for 1 minute. Let it dry for 24 hours.

NOTE: The color will appear darker when wet and will be several shades lighter after it has dried.

5 Wrap the hand-dyed twill tape around the gift using the instructions on page 16.

PAINTED WOODEN BEAD RIBBON ENDS

You can give plain ribbon a little pop by adding painted wooden beads to the ends. Wooden beads come in a variety of sizes and can be found at your local craft store. You can use acrylic paint, spray paint, or paint pens to paint the wood beads.

MATERIALS

Acrylic paint

Four ⅜-in-/1-cm-diameter wooden beads

1 piece of ribbon, 15 in/38 cm long and ½ in/12 mm wide

Wrapped gift

TOOLS

Paintbrush

Scissors

1. Using the paintbrush and the acrylic paint, paint the wooden beads. Let the paint dry for 24 hours.
2. Place the ribbon around the wrapped gift and tie it in a simple knot on top of the box. Trim the tails of the ribbon so they do not hang over the edge of the box.
3. String the painted wooden beads on the ribbon ends and secure with a double knot. Trim the ribbon ends.

Use twine or string instead of ribbon.

You can dye the wooden beads using the method discussed on page 115.

LINEN RIBBON

Linen is one of my favorite textiles because of its texture and simplicity. Linen ribbon tends to be pricey, but you can make your own using scraps of linen. Fraying the linen takes a bit of time, but the end result is a beautiful rustic ribbon.

MATERIALS

1 piece of linen, ½ by 23 in/ 12 mm by 58 cm

Wrapped gift

TOOLS

Sewing needle

BRIGHT IDEAS

Dye neutral linen a bright color using the method discussed on page 115.

Use fabric paint to create a pattern on the linen, as on page 112.

HOW TO

1. Starting on one short edge of the linen, use the sewing needle to loosen a crosswise thread, the top thread running parallel with the edge. Remove the entire thread. Once the crosswise thread is removed, the lengthwise threads will be revealed and will create a fringe.

2. Continue removing crosswise threads one by one until you have the desired amount of fringe.

3. Repeat steps 1 and 2 on the remaining three edges of the linen strip.

4. Wrap the linen ribbon around your wrapped gift and tie a bow according to the instructions on page 16.

CUSTOM PAPER ENVELOPE

MAKES

One envelope, 5¾ by 4½ in/ 14.5 by 11 cm (holds an A2 size card, 5½ by 4¼ in/ 14 by 10.5 cm)

MATERIALS

- 1 sheet printer paper, 8½ by 11 in/21.5 by 28 cm
- 1 sheet text-weight paper, 8½ by 11 in/21.5 by 28 cm, in your chosen color
- Glue stick
- 1 A2 card
- Sticker or washi tape (optional)

TOOLS

- Computer and printer
- Custom Paper Envelope Template
- Scissors
- Pencil
- Ruler
- Bone folder

Use patterned paper from your local craft or scrapbooking store to make envelopes.

Try printing a photograph on the text-weight paper and making an envelope out of it.

A custom envelope is a pretty way to personalize a gift. This envelope fits A2 cards, available at stationery stores, or you can make your own card. You can choose a beautiful solid-color paper, a graphic patterned print, or even found papers, such as newspaper, magazine or book pages, graph or tracing paper, or maps.

1. Download and print the envelope template on www.chroniclebooks.com/prettypackages. Using the scissors, cut along the solid exterior lines of the template.

2. With the pencil, trace the template onto the text-weight paper. With the help of the ruler, draw in the dashed lines from the template.

3. Cut out the envelope along the solid lines.

4. Using the ruler and the bone folder, score the envelope along the dotted lines.

5. Fold in the side flaps along the dotted lines. Fold the bottom flap up and glue to the side flaps.

6. Insert the card with your message into the envelope. Fold the top flap down and seal the envelope with glue, a sticker, or washi tape.

NAIL POLISHED BRADS

These colorful brads will add sweet details to your gifts. You can buy brads at any office supply or craft store and use them to close a bag or affix a tag to a gift. If you don't have nail polish handy, you can substitute acrylic paint or spray paint (make sure to apply a final layer of clear acrylic sealant to prevent chipping).

MATERIALS

Metal brads in assorted sizes
Nail polish in different colors
Clear nail polish top coat
Glitter nail polish (optional)

TOOLS

Piece of foam or cork, or masking tape roll

BRIGHT IDEAS

Use a paint pen or fine-tip brush to decorate the painted brad or write the first initial of the gift's recipient.

Cut out a 1-in/2.5-cm paper flower from a scrap piece of paper, punch a hole in the center, and place the painted brad through the hole to attach to a gift.

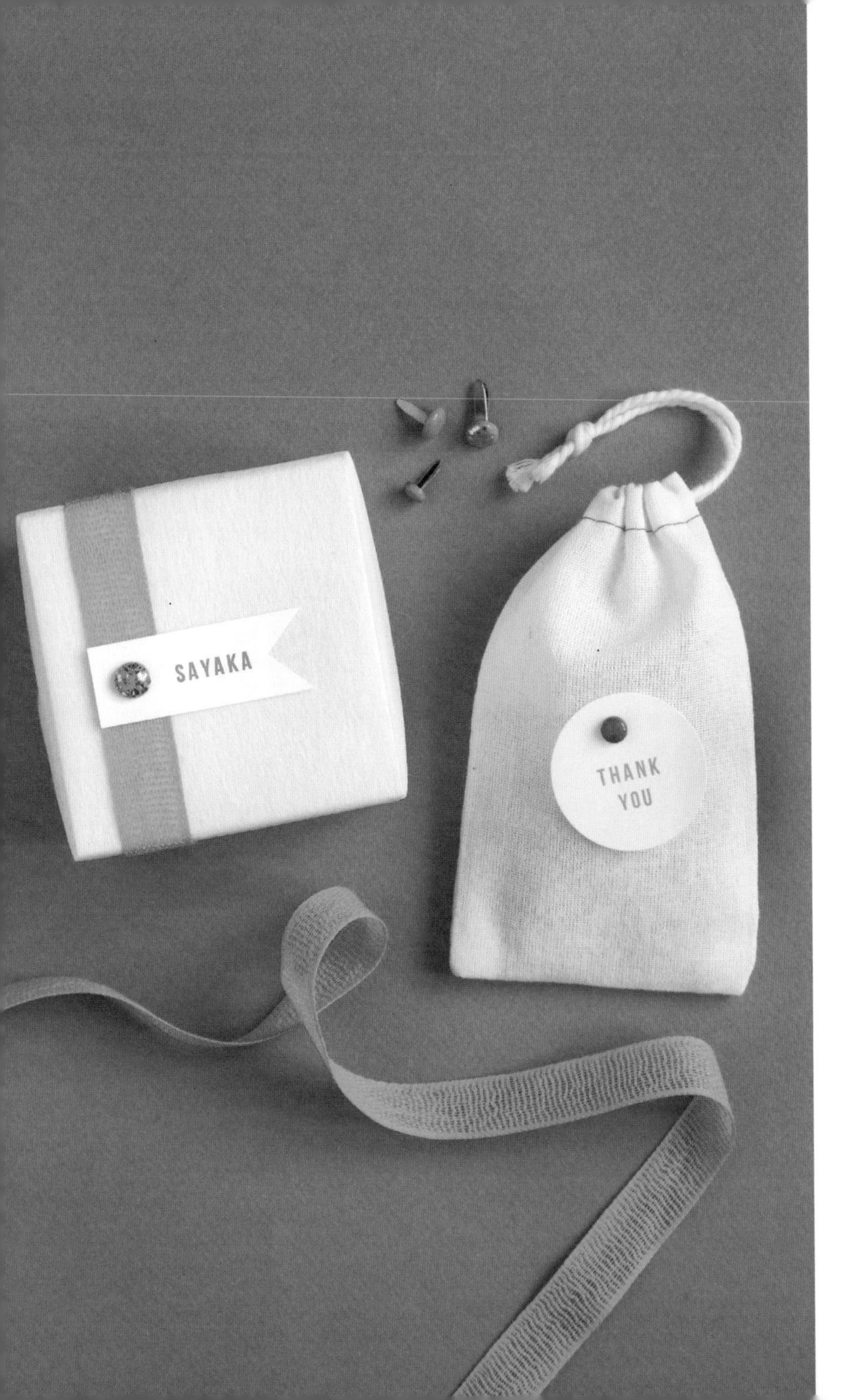

1. Press the metal brads into a piece of foam or cork. If you don't have either you can peel back a layer of masking tape on the roll, and adhere the brads to the roll by pressing the tape over the brads to secure them in place.

2. Apply two or three coats of colored nail polish on the heads of the brads. Let the nail polish dry for 15 minutes between coats.

3. Apply a layer of clear top coat on the brads to seal, and let dry for 30 minutes. If you like, you can add a layer of glitter nail polish instead of the clear top coat.

8

RESOURCES

ART AND CRAFT SUPPLIES

The following sources have stores in many U.S. cities and/or online mail-order sites.

DICK BLICK ART MATERIALS

Stock up on crafting basics and specialty items at Dick Blick Art Materials. You can find fabric paint, hand-carved stamp supplies, ink pads, paints, paper, pens, and adhesives. They also carry a large selection of tools, such as X-ACTO knives, rulers, cutting mats, and bone folders. *www.dickblick.com*

JO-ANN FABRIC AND CRAFT STORES

Jo-Ann stores carry fabric, felt, wool roving, ribbons, trims, string, twine, sewing needles, and lots of notions. You can also find a good selection of paper supplies, adhesives, and crafting tools. *www.joann.com*

MARTHA STEWART CRAFTS

This is one of my favorite sources for crafting supplies. Martha Stewart Crafts has a wonderful range of paper crafting tools, paper punches, paints, glitter, and packaging supplies. *www.marthastewartcrafts.com*

STAPLES

I love office supplies stores and Staples is my favorite. Pick up kraft or butcher paper for wrapping gifts, as well as miscellaneous embellishments, letter stickers, and label makers. *www.staples.com*

XYRON

This company carries terrific sticker-making machines and adhesives for crafting. *www.xyron.com*

PAPER

NASHVILLE WRAPS

Nashville Wraps has a good selection of solid-color and patterned tissue paper. They also sell boxes, bags, and various packaging supplies. *www.nashvillewraps.com*

PAPER AND MORE

My favorite resource for colored vellum paper. You can also find lightweight paper and cardstock in a variety of colors. *www.paperandmore.com*

PAPER MART

You can find all things paper-related at Paper Mart. They carry boxes, bags, kraft paper, butcher paper, newsprint, tissue paper, and more! *www.papermart.com*

PAPER SOURCE

If you love paper, you will love Paper Source. They carry an amazing selection of paper in various weights, wrapping paper, decorative paper, boxes, bags, and craft supplies. *www.paper-source.com*

PACKAGING SUPPLIES

BESOTTED BRAND

This is my favorite resource for rubber stamps. You can select one of their beautiful designs or have a rubber stamp custom made. They also carry a wide selection of packaging supplies and the best twine.
www.etsy.com/shop/besottedbrand

ULINE

My favorite place to buy packaging supplies in bulk. They offer a wide variety of butcher paper and kraft paper as well as gift boxes, paper bags, and glassine bags.
www.uline.com

RIBBON AND TRIM

STUDIO CARTA

Studio Carta carries the most beautiful Italian cotton ribbon and twine in a wonderful assortment of colors. They also offer crafting tools imported from Italy.
www.angelaliguori.com

TWILLTAPE.COM

This online store carries a variety of twill tape in various weights and widths. You can also order custom printed twill tape. *www.twilltape.com*

SPECIALTY SUPPLIES

CUTE TAPE

This online shop has the best selection of Japanese washi tape. They also carry packaging supplies, rubber stamps, stickers, and decorative tapes. *www.cutetape.com*

FRAMING SUPPLIES.COM

This is my favorite place to buy 3M Scotch ATG 700 Transfer Tape Dispensers and refills.
www.framingsupplies.com

TANDY LEATHER FACTORY

A great resource for leather and leather tools, including an alphabet stamp set and leather hole punch.
www.tandyleatherfactory.com

ACKNOWLEDGMENTS

The creation of this book was a collective effort on the part of many special individuals who I had the great fortune to work with. I always told myself that I would write a book if Chronicle Books came knocking on my door. And I am so happy that they did via my wonderful editor at Chronicle Books, Lisa Tauber. She gave me the opportunity to make *Pretty Packages* a reality and provided me with support, wisdom, and encouragement along the way.

Thank you to my dear friend Sayaka Chiba for loaning her beautiful hands and assisting me on the how-to photographs in the book. Her effortless style and grace always inspire me.

To my friend Joke Vande Gaer, who loves packaging as much as I do and who I have had the honor of collaborating with on packaging workshops and projects: Your encouraging words and advice always hit the spot.

And to all my friends and family members who checked in on me throughout the book-writing process, giving me pep talks and cheering me on, I could not have completed this book without you.

My deepest gratitude to my husband, Sang, and our two sons, Jeremiah and Judah. Our family of four brings me the greatest joy. I thank them for championing the book from the start and for being patient and understanding when I was working on deadline. Without their support and encouragement, this book would never have come to fruition.

HAPPY WRAPPING!
SAYAKA